W9-ATL-791

Also by Erich Hoyt

Seasons of the Whale

The Whale Watcher's Handbook

Orca: The Whale Called Killer

Riding with the Dolphins

Meeting the Whales

Extinction A–Z

The Whales of Canada

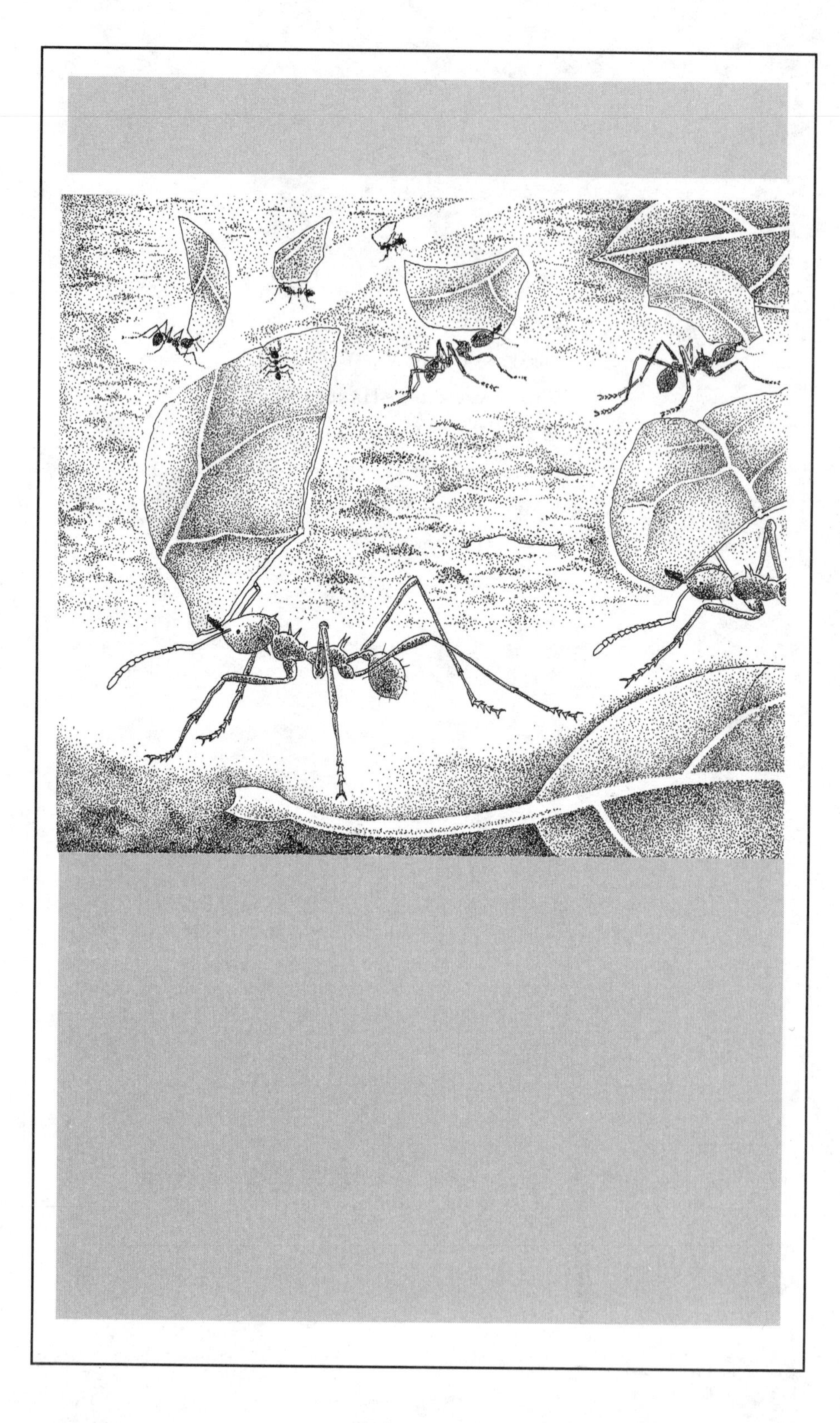

The Earth Dwellers

Adventures in the Land of Ants

Erich Hoyt

A Touchstone Book
Published by Simon & Schuster

Touchstone
Rockefeller Center
1230 Avenue of the Americas
New York, NY 10020

First Touchstone Edition 1997

Touchstone and colophon are registered trademarks
of Simon & Schuster Inc.

Designed by Barbara M. Bachman
Illustrations by Ruth Pollitt

Manufactured in the United States of America

1 3 5 7 9 10 8 6 4 2

Library of Congress Cataloging-in-Publication Data
Hoyt, Erich.
The earth dwellers: adventures in the land of ants / Erich Hoyt.
p. cm.
Includes bibliographical references and index.
1. Ants. 2. Leaf-cutting ants.
3. Ants—Research—Costa Rica—Estación Biológica La Selva.
4. Wilson, Edward Osborne, 1929– . 5. Brown, William L.
6. Myrmecologists—United States—Biography. I. Title.
QL568.F7H68 1996
595.79′6—dc20 95-45885
ISBN 0-684-81086-7
0-684-83045-0 (Pbk.)

To Sarah, Moses, Magdalen, and Jasmine,

with love and affection

Contents

Author's Note

My goal in this book has been to illuminate the daily dramas in the lives of ants and other insects, as well as of the scientists who approach them on bended knee. To do that, I have sought the perspective of viewing from less than an inch off the ground, as well as tunneling twenty feet below the earth and looking out from the inside of a hollow rain forest tree. In some cases, I have arranged details of events in insects' lives to make a concise narrative. However, everything in this book is, to the best of my ability, true and accurate.

The Earth Dwellers

Go to the ant, thou sluggard,
Consider her ways and be wise!

—Proverbs 6:6–8

Prologue

The idea for this book came to me one steamy morning a decade ago, deep in the Chaco region of Paraguay in the heart of South America. Driving the Pan American Highway on an expedition that never came to much scientifically but led to all sorts of things, I suddenly noticed a flashing green ribbon of life. We quickly stopped the car to investigate. Through most of the Chaco, this "highway" is a one-lane dirt road, in the rainy season a river of mud. This morning was sunny and hot, but we were already wheel-well-deep into the rainy season. The sun helped only to put a shiny veneer on the mud, forcing us to travel ever more slowly. At least we missed nothing, not even that green ribbon that turned out to be ants trudging across the road—a procession of leafcutter, or parasol, ants carrying pieces of leaves over their heads.

Stepping out of the car, we followed the parade on foot. A few hundred feet off the road, the ribbon led to a massive clearing on high ground where a series of mounds, turrets, and air holes marked the roof of a single, vast leafcutter colony. Ants! Ants coming and going, ants everywhere. We weren't looking for ants, but we had certainly found them. It was as if we had chanced upon an ancient civilization, some forgotten band of Mayans, untouched by time and still carrying on. I had heard about these impressive productions of the ants before, but in life they seemed much grander and stranger, more imposing. There were mushrooms growing out of the colony roof, and I suspected that these were the overgrown fungi the ants cultivate, fertilized by the leaves they collect, and then harvested to feed the queen and her extensive brood deep inside the colony. Peering down one of the holes, I wondered what went on inside that dark, mysterious, subterranean world. The scent of my breath immediately alerted them, and I was greeted by three oversized soldiers. They jumped out, their toothy mandibles snapping at me. I backed off, my curiosity satisfied for the moment.

Something struck me about the utter audacity of these animals to include the road, marginal as it was, as part of their foraging grounds. Yet the Chaco was more their world than ours. Ants clearly owned the Chaco. They could have it, I thought at the time; few areas have a climate so extreme and unforgiving that visitors regularly use the word "hell" when describing it. I was only dimly aware then of the extent to which these and other species of ants had transformed all manner of ecological niches, tropical and temperate, desert and grassland, nearly everywhere. Ants, in many ways, truly owned not just the Chaco but the world.

After South America, I looked at ants back home in the north temperate zone with a bit more curiosity and grudging respect, but mostly with ignorance. There were fierce wars being conducted on the sidewalks of the eastern United States. There were the carpenter ants eating through the walls of a house I lived in for a time. There were the fire ants that seemed to have conquered the American South, effectively preventing me from camping out on one cross-country vacation. I can't say I cared much for them. Then I was awarded a mid-career science journalism fellowship to the Massachusetts Institute of Technology (MIT). As a Vannevar Bush Fellow, I audited classes at MIT and Harvard University. One course I chose was Professor Edward O. Wilson's evolutionary biology course at Harvard. Wilson is among the premier ant scientists—myrmecologists—in the world. There was little talk of ants in class, but I was inspired by the reach of his knowledge. His ant studies had led to important findings in evolutionary biology, including population biology and genetics, ecology, sociobiology, and even philosophy. Week by week I began to see life as the evolutionary biologist sees it. At the same time, Wilson's occasional tales of fieldwork—his life among ants—were always delightful digressions, for me sometimes the high point of his lecture. I started staying after class and soon met with Wilson in his office. I wanted to ask him how a boyhood passion for the ants of Alabama, where he largely grew up, had blossomed into this impressive world view. What was so special about ants? He invited me to accompany him on his next trip to the tropical rain forest.

In March 1987, journeying to Costa Rica for several weeks, I joined Wilson and his old friend, mentor, and collaborator William L. Brown, Jr., from Cornell University. It was an ant-prospecting expedition, yet the talk stretched to almost every topic under the tropical sun. I began to see the exalted role that ants and other insects play on Earth, a role that reaches its most diverse expression in the tropical lowland rain forest. The stories they told of these ants and other insects—as part of a seeming morning ritual while lacing up their hiking boots or, later, over lunch or dinner—were the stuff of science fiction. Yet they were true.

As I began to read some of the extensive literature on ants and other social insects—much of it tucked away in small circulation journals or heavy tomes known mainly to specialists—ants became much, much more than just uninvited picnic guests. I learned about ants staging tournaments and all-out wars, sacrificing their lives for their queen and invading other ant nests, executing foreign queens, taking slaves. Some slave raiders are anatomically unsuited to gathering food, feeding themselves, raising their young; they *have* to take slaves. And I learned: Ants are stronger than elephants! The strongest elephant lifts less than its own weight, but ants, working together, can lift several times their own weight. Ants are essential to life on Earth. They do most of the Earth's turning and recycling of materials, building up the soil. Harvester ants specialize in seeds that they store in underground granaries while various tree ants plant hanging gardens in the treetops. Honey ants live on sweet juices, either from flowers or the honeydew from plant-sucking insects. Some ant species keep honeydew insects, such as aphids, as "cattle," feeding and taking care of them, and milking them as needed to nourish the colony. These and other symbiotic "guests" of ants include many species of beetles, flies, and mites collectively referred to as myrmecophiles—literally, "ant lovers." There are the leafcutter ants, confirmed fungus-eaters, or fungivores, loosely considered herbivores, as well as the seed-eating granivorous ants. Mostly, there are the carnivorous ants, ranging from skillful scavengers to hunter specialists in tiny insect prey to the army ants, described as the Huns and Tatars of the insect world; they hunt in massive regiments, overwhelming almost every other ant in their path as well as large and small insects, lizards, and even disabled birds and mammals. I began to realize that these stories that happen around the world and literally beneath our feet are unknown to most people or are widely misunderstood. I began talking to other myrmecologists and visiting entomological laboratories. I listened and I watched. I made more trips to rain forests, deserts, savannahs, temperate woodlands, and islands large and small to see how the many species of ants had met the challenges of diverse environments. There are about 9,500 named species of ants in 16 subfamilies and 300 genera, all of which belong to the family called Formicidae, the family of ants. The ants in turn belong to the order Hymenoptera (the bees, wasps, and ants), one of the most successful insect orders. I never cared much for insects and still don't like many of them, yet I have become partial to ants—a budding myrmecophile.

THIS BOOK WILL recount the everyday dramas in the lives of ants and other insects, interspersed with the stories of the myrmecologists who love to watch and study these small animals. We will follow three generations of a colony of leafcutter ants in the lowland tropical rain forest at La Selva Biological Station

(Estación Biológica La Selva) in Costa Rica. We will also meet many other colonies of ants in this insect-rich protected forest—including the legendary army ants with their million-strong swarming raids and the mysterious camouflage ants, considered the slowest, dirtiest ants in the world; their habit of not washing usually keeps them out of harm's way. And there are the tree-based aztec ant colonies, who battle each other for domination of the fast-growing, showy cecropia trees. Each of these colonies has a different strategy for survival, including sometimes bizarre behavior, clever tactics for getting their genes into the next generation. The activities of the scientists we'll see at work are sometimes as surprising as the ant behavior. Much of this book is situated in one part of the tropics, but I will not sacrifice the world view appropriate to the story of the ants and the myrmecologists. One piece of research in Costa Rica, or an account of an ant species found there, can lead to three stories of related ants in Madagascar, Mongolia, and the Swiss Alps.

I have two goals with this book:

First, to dip down to insect level in order to enter into the intimate workings of ant colonies—to present a true feel for what life is like from a perspective few but entomologists can imagine. It is not just a matter of reduced scale but one of a different perception. Ants and most other insects live in a chemical world. Smell and taste are their critical senses and means of communication. Sight, hearing, and touch are present to varying degrees but are less important. There is also the fundamentally social aspect to ant life. The insect societies can only roughly be compared to human organizations. Yet we can learn something about the nature of all societies and of social behavior by looking at ant colonies.

Second, to relate the tales of the myrmecologists themselves. If ever there was a warm and funny, yet strange and obsessive breed of scientist, these are they. Following the stories of past and present ant fanatics, I have tried to show how an apparently innocent interest in collecting as a child can get out of hand and how a predilection for very small fauna can expand to all of life. The colleagues of Wilson and Brown, myrmecologists present and past, represent distinguished scientists who have made their mark in human affairs as well as ant research. Ant study has led to solid advances in the study of evolution, social behavior, ecology, and chemical communication. These studies have absorbed some brilliant scientific minds and, to what seems an almost devious extent, have led to some grand theory-making over the years. At Harvard in the 1920s, Professor William Morton Wheeler, a socialist and one of the best known biologists of his day, saw the cooperation of ants as a good model for human society. Nearly fifty years later, ants led E. O. Wilson to his controversial ideas about sociobiology—the study of the evolutionary basis of social behavior—which he extends to include humans. Even today Wilson continues to think along these lines. And it will not be the last time that, with some justification, a big biological theory is fashioned on the backs of ants.

And now, as E. O. Wilson enjoys putting it, to the little things that run the world!

Chapter One

Workers' Paradise

A person, on first entering a tropical forest, is astonished at the labors of the ants: well-beaten paths branch off in every direction, on which an army of never-failing foragers may be seen going forth, and others returning, burdened with pieces of green leaf, often larger than their own bodies.

—*Charles Darwin*,

JOURNAL OF RESEARCHES INTO THE GEOLOGY AND NATURAL HISTORY OF THE VARIOUS COUNTRIES VISITED BY H.M.S. BEAGLE, *1839*

That story of . . . the gold-digging ants of India, which were as large as leopards, and whose hides were seen by Nearchus in the camp of Alexander the Great, and whose horns were mentioned by Pliny as hanging, even in his time, in the temple of Hercules —this and the many other ant stories invented or disseminated by ancient and modern writers are certainly not devoid of interest, but the actual behavior of the insect is so much more fascinating that you will pardon me for not dwelling on them.

—*William Morton Wheeler*,

SOCIAL LIFE AMONG THE INSECTS, *1923*

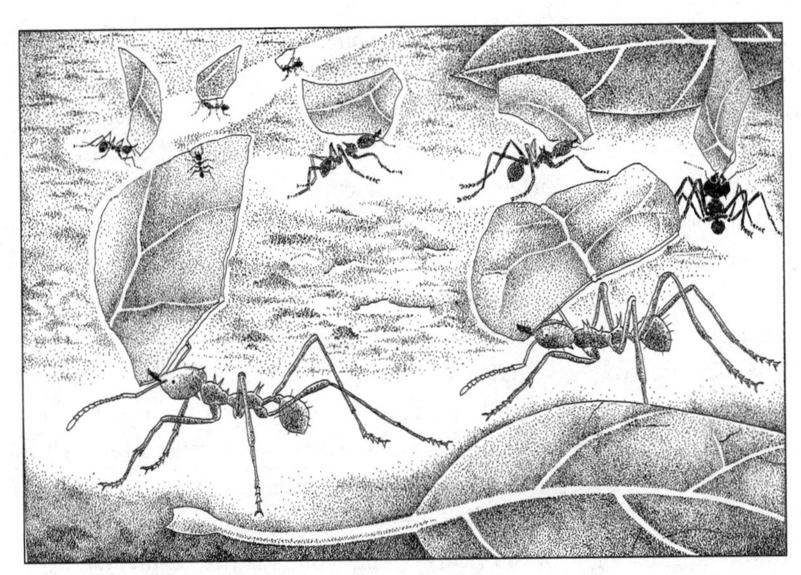

Rain Forest Highways. Leafcutter ants, Atta cephalotes, *carry their leaf burdens home through the jungle. Riding shotgun on the scout's leaf, one of the smallest ants in the colony helps keep watch.*

IN the steamy first light of the tropical rain forest, at La Selva Biological Station in Costa Rica, life is buzzing. It is February, nearly the season of new beginnings for many species. The forest echoes with the whooping calls of those gregarious blackbirds, the oropendolas. At lower volume but ever more persistent is the hissing of thousands of cicadas and other insects. And in the distance: howler monkeys on dawn patrol.

No matter the racket, nocturnal animals such as the silky anteater and the common vampire bat are winding up their day and preparing to sleep, while certain birds and others that need the light to hunt are just rolling awake and looking for breakfast. Still others—including some of the insects—have been working all night and will continue into the day.

From overhead, slanting through a break in the dense forest canopy of strangler figs, lianas, and other vines, a shaft of light beams down on a scene of frenetic activity. A worker ant—a foraging scout, female, as are all ant workers—stands on the leaf of a low-growing bush beside a fallen tree. Her hind legs are digging in, her head is down, hard at work. Her serrated mandibles, her jaws, are moving, sawing through a leaf. The air is pungent with leaf sap. As it drips from the leaf, she stops to lick a drop or two for refreshment. Two seconds later she's back at work.

As she saws, the sharp blade of her mandibles cuts through the leaf tissues, and her round head bobs up and down with the rhythm of a woodcutter. Up, down, up, down, and slowly she begins to pivot on her hind legs to inscribe a curved line. She turns steadily, skirting the midrib of the leaf and finally coming all around in a nearly perfect circle. The cutout complete, she nudges the leaf panel to break the final perforations, and it falls to the ground. She jumps down, squats over the leaf piece, and marks it with a drop of liquid from the underside of her rear abdomen, her gaster. This marking will make the piece more attractive to the others in the colony, once she returns to the nest. It may also contribute enzymes that will eventually help break down the leaf. Then she grabs the leaf piece with her mandibles, hoisting it high so that it almost rests on her antennae like a floppy, oversized hat. She walks a few feet into the full sunlight of the forest clearing and stops. Her leafy hat, her green parasol, is already starting to fall off. It's heavy.

In the speckled light of the clearing, the scout's entire segmented body, seen in profile from head to gaster, glows reddish brown—except for the protruding black eyes. She has the distinctive spines of a leafcutter, or parasol, ant. They look like thorns rising from her back. She is covered in fine, tiny hairs from her head to the rear feet, or tarsal claws. The two antennae droop under the weight of her green load—the piece of leaf is much longer than the ant herself. Now she steadies the parasol once again, splaying her six sturdy legs as she grips the leaf firmly in her jaws and hefts the oversized cargo. The leaf fragment is three times her body mass. Leafcutters typically carry in a single load one and one-half to six times their body mass. She struggles to press the green prize up high. It takes another second or two to distribute the weight. This time it looks more like a green flag than a parasol. And then, threatening to break the camel's back, another, much smaller leafcutter ant appears out of nowhere, climbs aboard the leaf piece, and clings to one end of it, riding shotgun. She is less than a quarter the size of the scout, her nestmate. This smaller worker—a sometime companion to the scout—keeps alert for enemies, listening for the annoying buzz of the flies that parasitize many species of ants with their eggs. She, too, marks the leaf piece with a drop or two from her abdomen, contributing more enzymes to help break down the leaf.

Today the scout has what may well be a new leaf species—an untried plant for her colony. Her young, tender leaf-prize thrives on a few low-lying bushes at the edge of a clearing, some distance from the nest. It is an exotic plant—an introduced (that is, not native) species, perhaps a weedy plant that escaped from a nearby farmer's field. As an exotic, it has fewer of the built-in defenses that many indigenous plants have evolved through living among insects for

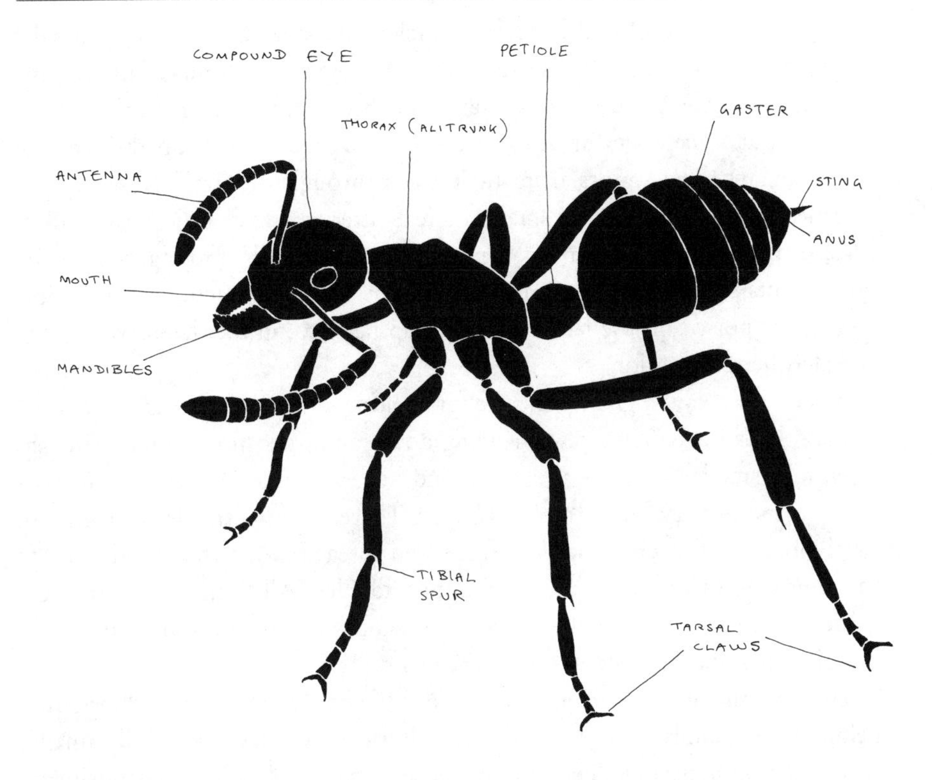

The Parts of an Ant

millions of years. Therefore, herbivorous insects, mammals, and other animals may be able to make easier use of it.

In a survey of forty-two plant species from the dry forests of Costa Rica, 75 percent contained terpenoids, steroids, waxes, and other defense compounds that repel leafcutters. Many tropical forest leaves are tough or have sharp spines that discourage monkeys and other animals from eating them—even more than chemical defense compounds. But the persistent leafcutters aren't much bothered. Their razor-sharp mandibles can slice through some of the fleshiest leaves and even twigs and stems. And with regard to the defense compounds, the leafcutters are selective about what they cut, but in species-rich tropical forests they have a lot of choice. Surprisingly, the rarity or commonness of the plant species has no real bearing on its appeal to the prospective leafcutter scout. Neither is its proximity a crucial factor; leafcutters will walk a few hundred feet to a favored plant species. In all, researchers estimate that the attine tribe of gardening ants, which leafcutters belong to, cuts 12 percent to 17 percent of the leaves and flowers produced in the New World tropical forests they inhabit. On Barro Colorado Island in Panama, leafcutters annually cut almost 250 pounds of foliage per acre. Or, to put it in terms of the colony, every day a single mature colony uses as much vegetation as a mature cow.

The scout sweeps her antennae from side to side to pick up the scent of her nestmates. The vapors from their trails through the jungle often linger. Myrmecologists describe the smell of a leafcutter trail as faintly grassy, sulfurous, or fruitlike, with a hint of naphtha. But: she smells nothing. Still, many other scents persist upwind—smells of danger. The tiny leafcutter riding shotgun stands poised on the leaf, ready to snap her jaws in the defensive posture and alert her companion.

A few feet away, a giant ponerine, or hunting ant, stands at rigid attention, on patrol, its huge, lethal jaws hanging down almost to the ground. This ant often kills smaller ants like this scout and her even tinier associate, and feeds them to the hungry larvae of her colony. There are also smells belonging to carnivorous beetles, crafty spiders, frogs, and anteaters. And the sounds include the annoying buzz from the parasitic phorid flies. All spell danger to these gardening ants in their daily activities of scouting for new leaf sources and bringing them back to the nest for processing.

The flies, the most irritating, hover above the hardworking ants. Phorid flies belong to the family of small flies called Phoridae, which are generally smaller than common houseflies. They follow the ants' columns, trying to deposit eggs on unsuspecting workers. The scout is helpless because she is carrying the leaf, but the shotgun ant now comes in handy. She takes a few swipes at the flies with her front legs, antennae, and mandibles, rearing up on her middle and hind legs, though the flies are not much deterred. Phorid flies bother many species of ants, and even the fiercest ponerines can be driven to distraction, ducking their heads and jumping around to avoid the nuisance flies.

Because the scout has left the main trunk line, or path, that the colony uses, she crooks her head to the sky. Some species of ants, as well as other insects, are able to orient visually. They can remember patterns of overhead branches and leaves in the canopy as well as large landmarks such as rocks or bushes, and use them to find their way. Certain ants, bees, and other insects are also capable of celestial navigation and use patterns of polarized light in the sky to get back to their nests.

As she runs, she lays a trail—a liquid chemical squirted out from the poison gland in her lower abdomen—in case her leaf proves valuable. Other colony workers can then use this chemical trail to find the tree or bushes and strip the area. The chemical, or pheromone, is a glandular secretion that many insects such as ants use to communicate within their own species. Though long suspected, this was only proven in the late 1950s after E. O. Wilson began his ant pheromone research at Harvard. Most of the chemicals are produced in exocrine glands in the lower abdomen, or gaster, but some are in the head, especially near the mandibles. Ants can taste the odors on their tongues, but mostly they pick them up in vapor form through more than two hundred cones that function as odor receptors on their antennae.

Depending on the ant species, some pheromones sound an alarm or, like perfumes, are requests for grooming. Others say "feed me," "follow me," "help me," or "let's cluster together and dig." The pheromones for marking trails are sometimes understood by other ant and insect species, who may use some of the same trails or wait beside them to try to pick up food or other material dropped by the ants. Certain beetles, like highwaymen, wait to try to rob the ants of their food by giving them the ant's own "feed me" scent. Still other pheromones are used for territorial or home range signals or nest markers. From smell and taste alone, an ant knows her nest and can recognize other castes and the mother queen. With all the ants programmed to respond to chemical messages and perform their roles, it's a real demonstration of better living through chemistry. Most pheromones are so powerful they evoke immediate responses in ants. They can also be enduring. The trail markings of certain leafcutters can last for six to twelve days; with army ants, the trails can last a month or more, particularly in the dry season. One researcher documented an army ant trail some thirty-one days old.

The scout worker and her cohort hurry on toward the main trunk line where sister ants—other foragers who have been out for the morning—are returning to the nest. Soon she encounters the familiar scent of the trunk line. Up ahead, the ants are moving along a three-lane dirt highway that they have cut through the jungle and have maintained for many months. The smell of the highway lures her and tells her that she is almost back on the road that leads home. She stops a worker so the two can wave antennae over each other to verify that they are from the same colony and caste, but also to tell her sister, through the chemical in the pheromone, that this is a new leaf she has found.

At the first break in traffic, the scout butts in, heading back to the colony. At once the pace picks up. The trail vapors are clear and strong now. She is running but has only to follow the lead of the ant ahead of her, carrying a piece of a leaf from another plant, who is following another ant farther ahead that is also loaded down, and so on. The leaves—carried above the oversize heads of so many ants all in a line—resemble an ant Easter parade, the strollers sporting many odd-shaped parasols, source of one of the leafcutter's nicknames. To the ants, the parasol is pure accident; it's only the human view from above. And many of the parasols have the tiny ants riding shotgun on top of the leaves. In the long shadows of morning, in any case, the big-headed ants receive only flickering protection from the sun. Their comparatively large heads led in part to the scientific name of these particular leafcutters, *Atta cephalotes;* the root *cephal* comes from the Greek word for "head." Soon, however, they will duck inside to the coolness of the subterranean nest. That is the goal for these foraging workers—bringing the leaf bounty back home to their nest at the edge of the forest near the overgrown plantation along the Río Puerto Viejo. The long line of leafcutters now extends for hundreds of yards through this forest, along this ant highway swept clear of all debris. Two lanes, a regular speed lane and a passing lane, lead toward the colony nest, while the third lane is for ants venturing out from the nest to cut more leaves.

LEAFCUTTER ANTS are considered one of the main herbivores throughout the Americas. They are not properly herbivores, however. They don't eat the leaves they cut—many contain poisons—though the workers occasionally take liquid nourishment from the sap. Instead, they feed the leaves to a certain species of mushroom, a fungus, that they cultivate in underground chambers. This is the source of another of their nicknames: fungus ants. *Atta cephalotes* grow, eat, *subsist* on fungus.

Of the approximately two hundred species of gardening ants, only thirty-five are properly leafcutters—cutting leaves, flowers, stems, or other living plant materials. The most advanced and social—with their great underground civilizations and bold earthmoving ventures—are several *Atta* species from the Latin American tropics. Besides the various leafcutters, other gardening ants range through Texas and some as far north as Long Island, New York, where they have been sighted working among Twinkie wrappers and the odd Coke can in abandoned tracts beside shopping centers and under power lines. These northern species are the least conspicuous members of the gardening tribe, but they, too, are highly evolved, or "derived," as evolutionary biologists put it. Instead of cutting large leaf pieces, they collect the corpses of ants and other insects, seeds, insect droppings, flower stamens, and other organic detritus to feed the precious fungus gardens. As do all other gardening ants, they deposit their own fecal droplets on the garden to help the fungus in digesting its food. The badge of this tribe is its extraordinary ability to cultivate fungi and to keep them alive and thriving. These ants are farmers, and very successful ones.

The identity of the fungi grown by gardening ants, however, remains a mystery. These fungi rarely produce fruiting bodies that are used for identification. The few fruiting bodies obtained indicate that the leafcutters are farming various forms or species of the basidiomycete family called *Lepiotaceae.* Some researchers believe that all gardening ants cultivate forms of a single fungus species, *Leucocoprinus gongylophorus.*

To feed their fungus gardens, leafcutters working long shifts in the Latin American jungles and savannahs will dismantle a tree, large bush, or other vegetation piece by piece. In some places they are serious agricultural pests, but a number of scientists come to their passionate defense. American myrmecologist Neal A. Weber lived and worked among the leafcutters and other gardening ants from 1933 to 1972. He praises ants as "one of the few animals that transfer soil mineral nutrients essential for plant growth to the upper layers where the nutrients can be utilized. By their tunneling, the ants also aerate the soil and create passageways for drainage or water penetration." Weber singles out leafcutters as "the local animal agent chiefly responsible for taking organic matter down into the soil in tropical forests. . . . Moreover, [leafcutters] accelerate this process by which such plant material is broken down to allow reuse of the chemicals it contains. The extremely dense mats of fine roots growing in the refuse deposits outside [their] nests attest to the abundance of nutrients in discarded substrate from the gardens." It all adds up to recycling—an essential service that ants perform worldwide and that the leafcutters excel at. Along with termites, ants do most of the turning of the earth—moving more soil than earthworms and much more than all the world's human farmers.

When the leafcutter ant workers are active aboveground, they are unmistakable, impossible to miss. Racing along on their network of highways through the jungle, they are in their element. Because of their prominence, there are many common names for them in the countries and cultures of the Americas: "wee-wee" to Nicaraguans and Hondurans; "hormiga arriera" to Colombians, Panamanians, and Mexicans; "saúva de mata" to Brazilians; "zompopos" to Guatemalans, Salvadorans, and Costa Ricans; and at least fifteen other Spanish, Portuguese, and native Indian names. Among Latin farmers they evoke a mixture of fear, dread, and respect. A motorist driving on the Pan American Highway can sometimes spot the ribbon of green along the road or crossing it. It can be seen during the day or even picked up in the headlights at night. If the motorist stops to investigate and gets down on hands and knees, the green ribbon becomes the many individual ants carrying their leaf prizes home. In daylight, from a helicopter hovering low over a scrub forest or a clearing in the rain forest, it is sometimes possible to see a dozen of these ribbons leading out in every direction, giving the appearance of a giant green asterisk above the bleached red or brown dirt background. The ribbons, which can extend for a hundred yards or more, always lead back to the leafcutter ants' nest.

The roof of the colony nest is 20 feet or more across, and some nest roofs extend up to 52 feet by 52 feet, an area more than 2,700 square feet. It often forms the centerpiece of a jungle clearing. The color of the nest roof differs from the surrounding soil because it has been unearthed from sometimes twenty feet below the surface. The roof has a built-up area with piles of sculpted dirt similar to sand castles on a beach. The raised area forces the sometimes heavy rains to run off. All around and on top of it are hundreds of openings—sometimes more than a thousand. Around the perimeter, nest entrances function as air intake holes while the openings in the roof are chimneylike ventilation holes. Several million ants busy at work down below, added to the heat of the fungus metabolism, could cause overheating and produce a foul, oxygen-poor atmosphere. But as the fungus cooks, the air in the central zone of the nest becomes hot and rises, pushing out the air from the central galleries. At the same time, stale air is drawn from the perimeter, through the tunnels, and up and out—creating better circulation and moderating the temperature and humidity of the nest.

APPROACHING THE NEST at the edge of the forest, the scout ant feels the cooling breezes off the river. She is following the long line of foragers that splits up and enters one or another of the entrance holes. At the door she almost trips over one of the bulky, big-jawed soldiers. The soldiers are tough, lethargic animals that mostly hang around waiting for something to happen. They are useless as workers. They do little unless there is an attack, but then they fight to the death.

She lets two soldiers wave their antennae over her to confirm her membership in the colony. Her leaf load just fits through the entrance. Down, down she goes, with many other foragers all in a line, through several feet of ever darker, more winding tunnels, the smell of the nest filling them. The scout puts her leaf booty on the floor of the first near-empty chamber she finds, and the tiny ant riding shotgun slides off the leaf. This chamber, with its rounded ceiling, is about the size of a human's head. Room sizes range from approximately five by eight inches to twelve by twenty inches. In the center of the chamber floor, the scout finds the fresh white fungal mycelium, the threads of the fungus, just beginning to spread. Her job is complete for the moment. It will be up to other, more specialized workers to find out if this leaf provides good food for growing fungi, whether the fungus accepts the food.

Before the scout departs, she licks a worker who has come to the chamber to examine the message about the new leaf. It tells the worker where to find the trail. But this worker in the chamber has other duties. Slightly smaller than the scout, she licks the leaf, scraping it clean. Then she cuts it up into tiny pieces, each about one-sixteenth of an inch across—about the width of a grain of rice. Finally, she and the other shredders step aside as yet smaller workers arrive—the pellet-makers.

The pellet-makers chew the edges of the pieces, crumpling and molding them into moist pellets. One by one the workers slip the pellets into the pile of fungus, covering with tufts or strands of fungus any that are exposed.

This is the underground garden of the leafcutter ants. From time to time other workers come into the room to add fecal droplets to the fungus. The fungus, the flocculent mycelium, looks like a grayish white cleaning sponge. It is alive and in the process of spreading from a few inches to a foot or two across. Even before it has filled the room, pieces of it will be stripped off to start gardens in other rooms. As the leafcutters from the various castes keep busy feeding and caring for the fungus, the question arises: Who is in charge here, who is keeping whom—the leafcutters or the fungus?

For naturalists and other New World explorers of the past five centuries, finding such an underground mass of fungi must have been like coming upon an ancient buried inscription in a language that they didn't understand. From the surface all they saw were the leaf-carriers, and this led to the assumption that the ants were eating the leaves or possibly—according to British naturalist Henry W. Bates in 1862—using the leaves to thatch a roof over the entrances of their colonies to protect the young from heavy tropical rains. In 1874, however, Thomas Belt, a self-taught British ecologist whose work as a mine geologist and engineer took him to Central America, reported his startling discovery: Tunneling down into the underground nests, he observed that the leaves are cut up and wedged in among what can only be called mushroom gardens. The leaves are simply the principal source of manure for what he called the "ant-food."

There may be some two thousand rooms in a mature nest, more than three hundred of them occupied by gardens, as well as a single queen, brood, and 3 million to 7 million workers. As each garden grows, the thick, cottony fungus spreads, reaching to the seven-inch-high or higher roofs of some chambers. New chambers must constantly be excavated and fresh gardens started. Throughout the colony's numerous channels, there are even smaller workers —the transplanters—responsible for taking fungus from chambers where it is thriving and moving it to chambers where new leaf pellets are ready. Within twenty-four hours the new pellet chambers will be mostly covered with fresh fungi ready for harvesting. Instant fungi!

The smallest workers of all—only one-sixteenth to one-eighth of an inch long and able to pass through the narrow channels deep inside each garden chamber—do the weeding, pulling out alien species of mold. These harvesters

—which myrmecologists sometimes call minimas—are one-three-hundredth the weight of a big soldier ant, yet they are the most energetic and the hardest-working ants in the leafcutter colony. The harvesters also share with all the other ants of the colony the nonstop job of cleaning. Ants have the ability to clean their nests and themselves with antibacterial secretions, which they spread around with their legs and mouth parts to combat the growth of bacteria and unwanted fungi—crucial for keeping their colonies clean and healthy. Only the good fungus—the species they live on—is cultured and allowed to survive.

Once the fungus is "ripe" and ready to eat, the harvesters pluck the minute, inflated tips, which Alfred Möller called kohlrabi because of its resemblance to heads of stalked cabbage. Working in southern Brazil in the 1890s, Möller became the first to see the ants eating the fungus. A little more than a decade later, shortly after William Morton Wheeler, the father of modern American ant studies, began his exhaustive studies of ants, including leafcutters, he renamed the kohlrabi *gongylidia*—the name commonly used today. Each gongylidium is bite-sized to an ant. The harvesters carry the food out to members of the colony and pass it around. Various workers then make sure the larvae get their share, though most of them are surrounded by fungi and have only to open their mouths. Larval ants must eat constantly: They are growing fast. From egg to adult, ants receive meticulous care—comparable to the attention mammals such as humans give to their offspring. But among ants the sisters—not the parents—do the caring. And the sisters also take care of their mother.

The job of feeding the queen is left mainly to the pellet-makers. Some of these workers are always with her, attending her in the special chamber where she resides. The colony mother and founder, she is a queen and is treated like one. She receives fresh gongylidia as well as a supply of large eggs that the workers lay to feed her. They wait on her. They groom her, licking and touching her. About the size of a baby mouse, she is an egg machine, laying an egg every two minutes, some twenty-seven eggs an hour. The workers carry them to the nurseries. From her station deep in the nest, the queen sends out a pheromone announcing her continued presence to all the colony. By being licked and touched by her pellet-maker attendants, she probably receives updated information on general colony health. Repairs may be needed in one part of the colony; workers may be trapped and need help in a collapsed chamber; more leaves may be needed here or there for fertilizer. The queen does not mediate decisions about any of these situations but probably helps pass the information around to be dealt with by the appropriate caste of ants. Workers responsible for the care and feeding of larvae, for example, can respond to information about the health of the colony by producing new types of workers as needed. There are at least seven divisions, or castes, of leafcutter workers, each with its own size class. When the colony loses workers of one caste in an accident or a battle, the colony must replace them. Other workers are called to cover for the missing caste while new broods of that type of worker are raised. All the eggs that the queen lays are alike. As the workers feed the female larvae different amounts of food and perhaps expose them to varying temperatures, the larvae grow into various sizes of workers—different castes. The colony is thus able to control the type of brood that is produced. The feedback loop from time of conception to replacement of a caste can be as little as a few weeks.

LIFE IS WORK and work is life to leafcutters and other ants. The seven castes of leafcutters perform twenty-nine different tasks or social functions, all in assembly-line fashion, everything from cutting the leaves to cultivating the gardens between the rows. In each caste, workers handle two or three different jobs suited to their size and physical ability. In laboratory research at Harvard, E. O. Wilson discovered that when new leafcutter ant colonies are founded, the young mother queen handles all the tasks, stripped down to the essentials, until she has raised a brood of workers. And later on, if one caste is removed experimentally, ants from other castes eagerly take on more work. It is the ideal assembly line when workers can fill in for each other as needed. Henry Ford would have hired them on the spot. Those who research ergonomics—the study of work, performance, and efficiency—in insects as well as other animal species are impressed by ants.

Ants, unlike social mammals, never play. But their reputation as the hardest working animal on Earth may not be deserved. At any given moment, only a small fraction of a colony's workers are actually working. Inside the nest, most are standing still, grooming themselves or just walking around aimlessly. The individuals of one typical ant colony were found to be inactive 78 percent of the time. In fact, as Bert Hölldobler and E. O. Wilson have written in *The Ants,* ant societies are about as industrious—or lazy—as human societies, although "lazy" may not be the right word to use. Their resting—their inactivity—may have survival value. By working at prime times and using the workforce only as needed, the overall energy of the colony is metered and carefully conserved.

The work of leafcutter colonies revolves around the fungus gardens—growing them, fertilizing them, harvesting them. Leafcutter ants capitalized on one of the great lucky breaks of evolution. It happened somewhere in South America during that continent's millions of years of geological isolation from the rest of the world. In what has been described by researchers as "an extremely rare event," the primitive ancestor of modern gardening ants captured a fungus and began feeding it and harvesting it before it developed spores to reproduce itself.

Back then, gardening ants probably came in only one size and lived the life of generalists. The colonies were likely small, and they used droplets of their own feces and seeds, insect droppings, and other organic detritus to fertilize the gardens, to feed the fungi. Over millions of years the ant-fungi relationship became mutually dependent, or symbiotic. In time the fungus lost its ability to produce the spores essential for reproduction. Subdued by the ants and unable to reproduce, the fungus came to depend on the ants to stay alive and propagate —just as the ants depended on it for their food. This event is thought to have occurred only once in the evolutionary history of the ants. In any case, the gardening ants took advantage of it and in time adapted their caste system and underground living quarters to growing, harvesting, and protecting the fungus. In this way, these ants managed the remarkable transition from hunter-gatherers to farmers—a rare status among ants. Most ants are either carnivorous scavengers or fierce hunters that prey on the adults and young of other ants, termites, beetles, mites, and many other arthropods, or invertebrates. Of course, leafcutter ants retain some aspects of the hunter-gatherer existence, sending out some of their workers to scout for new fertilizer sources, but most of the specially evolved caste members focus on farming. Only a few termites and certain wood-boring beetles share the sophisticated habit of culturing and eating fungi, but they use *dead* vegetation to grow their fungi. Of all animals everywhere, only the various species of leafcutters have evolved the ability to use fresh vegetation to grow fungi.

As simple underground "garden plots" gave way to vast subterranean farming cooperatives, leafcutter ants became the main herbivorous animals of the tropical Americas. This happened long before humans came on the scene. Higher social insects first appeared in their earliest forms about 200 million years ago. Beginning about 100 million years ago, during the mid-Cretaceous period when dinosaurs supposedly "ruled" the Earth, early social ants began spreading out and dividing into new species.

Sometime after the isolation of South America and the start of the Tertiary period sixty-five million years ago, primitive gardening ants began their intimate relationship with a fungus. The best estimate is that it happened about 50 million years ago. In any case, the most evolved gardening ants today have clonally propagated the same fungal lineage for at least 23 million years, according to Ignacio H. Chapela, Stephen A. Rehner, Ted R. Schultz, and Ulrich G. Mueller, variously from Cornell University and the U.S. Department of Agriculture in Beltsville, Maryland. In their 1994 *Science* paper entitled "Evolutionary History of the Symbiosis Between Fungus-Growing Ants and Their Fungi," they wrote that both the ants and their fungus have evolved over millions of years, but every leafcutter colony today is propagating fungus that can be traced back to its ancient origins with the same leafcutter species.

Shortly after the rise of the isthmus of Panama, some 3 to 8 million years ago, leafcutters spread north into Costa Rica, through Mexico and on into the United States. They were well established by the time the first humans arrived in the Americas from Asia. Following the development of human agriculture some ten thousand years ago in Mexico, the leafcutter ants became one of the leading agricultural pests in the New World tropics, raiding corn and bean patches by day and by night and carrying their loot underground. Leafcutter ants consume more vegetation than any other group of animals—including caterpillars, grasshoppers, birds, and even mammals. They currently inflict an estimated billion dollars or more of crop damage each year, mainly in tropical Latin America. The approximately two hundred gardening ant species today belong to the Attini ant tribe; they vary in size, number of castes, and preferred habitats. They are found only in the New World, in every country of the Americas except Canada and Chile. A few leafcutter ant colonies with young queens have been transplanted to butterfly parks and zoos in Europe. They are fascinating to watch, but should inseminated queens ever escape or be carried to tropical Africa, the result could be a catastrophic biological invasion—one from which Africa might never recover. To date, however, no leafcutters have bred in captivity. The colonies are kept small enough that they will not produce virgin queens and males.

Leafcutters were first brought back to Europe as specimens and listed in 1758 by the Swedish naturalist and physician Carl von Linné, better known as Linnaeus, who was then developing the modern system of naming species. In the early 1830s, Charles Darwin met leafcutters on his extended tour of the American tropics, part of the five-year H.M.S. *Beagle* expedition, and was later inspired to develop the idea of kin selection to explain the ants' altruistic societies with sterile worker castes. Various other European naturalists and scientists worked on leafcutters in the late nineteenth century—including Thomas Belt and Alfred Möller, as well as a number of Brazilians and the German-born Hermann von Ihering.

Leafcutters were the first ants to move William M. Wheeler to take up his studies in 1899 at the University of Texas. "I happened to see a file of [ants], each with its piece of leaf poised in its mandibles," he wrote later. "I vividly remember the thrill of delightful fascination with which I watched the red-brown creatures trudging along under their green loads, and it seemed to me that I had at last found a group of organisms that would repay no end of study." At the time there was no active myrmecologist in the United States, but two of the European fathers of myrmecology who had laid the foundations of leafcutter and other ant systematics, the Swiss Auguste Forel and the Italian Carlo Emery, helped Wheeler get started and sent him papers and encouragement. Thus was the modern era of American myrmecology born. Wheeler synthesized the early papers and observations on leafcutters in a 1907 monograph and soon went on to the American Museum of Natural History, and from there to Harvard. He became one of the most widely known and respected biologists of his day, initiating a distinguished line of Harvard entomologists, that continues to this day through the late Frank M. Carpenter to Edward O. Wilson. Wheeler's last student, Neal A. Weber, inspired by his teacher, devoted his life to studying leafcutters, and new secrets of the leafcutters are still being unraveled today by young researchers such as Ted R. Schultz of the Smithsonian Institution and James K. Wetterer at Wheaton College. Today they call themselves "attinologists"—myrmecologists who specialize in studying gardening ants—that is, "attines," or members of the ant tribe Attini. Their work is directly traceable to Wheeler's thrill at that line of leafcutters on a road somewhere near Austin, Texas, a century ago.

A FEW DAYS after the scout returns to the colony with the new leaf, the colony begins producing an excellent crop of fungi. Meantime, the foragers have been working to harvest the low-lying bushes where these leaves grow. Some four thousand ants have been mobilized.

After the early afternoon rains, the number of foragers departing for the site rises to six thousand. As they split from the trunk line, the pheromone on the new trail, left by the scout and refreshed by many passing workers, directs them on their journey. A hundred smells seep up from the moist earth, yet the message is clear.

One by one the ants take their positions on the long leaves and start the task of cutting out the materials, lowering them to the ground, and carrying them back to the nest. As one ant finishes on one part of a leaf, another climbs aboard to secure the next piece. Each leaf is like a patchwork quilt—and only the labor of thousands of ants harnessed in this way will get this job done. Even so, as more laborers become available, forty-five thousand more ants depart for the site before dusk.

Then, out of nowhere, a giant tropical ant, *Paraponera clavata*, looms on a branch overhead. In the jungle twilight, the creature's body glistens a shiny, coal black. The Latin Americans call it "bala," which means bullet, partly because of its hefty body, nearly the size of a .22 bullet, and because its sting spells instant death for other insects and considerable pain for humans or other mammals, although it is not deadly for any creature much larger than itself. Growing to nearly an inch long, it is the largest tropical ant—easily twice as long as the big leafcutter soldier. Tarsal claws on its feet enable the bala to scamper up trees. Its big-blade mandibles and overpowering sting allow it to hunt most ants, beetles, termites, and various insect larvae. Usually the bala does not bother the leafcutter ants. In fact, though capable of carnage, it is mainly a scavenger, and it also has a serious sweet tooth, searching out plant nectar, golden droplets of which it can carry in its mandibles back to the colony. Yet almost any ant that gets in its way is fair game and can be dispatched at once.

The leafcutter workers continue their jobs. The bala is too far away to be seen and can only be smelled faintly. No leafcutter worker even returns to the nest to recruit a soldier. There seems to be no immediate need to defend the colony's interest.

Then five more balas apppear. They are excited, making sharp clicking sounds with their mandibles and exuding a strange perfume. All of them run toward the leafcutter workers carrying their leaf loads. Some of the ants, partly obscured by leaves, do not even see their attackers, but they are alerted by the sudden acrid smells of the beasts bearing down on them. Then they see the blades of the mandibles as they snip them in two, crushing the head, severing or mangling limbs, crunching exoskeletons. Almost ten at a time, a hundred leafcutter ants are cut down. A few others are stung. A dozen more bullet ants loom into view and join the fray. A thousand leafcutter workers are killed.

One wounded leafcutter worker escapes, limps back to the main trunk line, orienting by the trees and the late day sky. All the way she lays an alarm-recruitment trail as thick as she can. Her rear legs are mangled, but the front four pull her along. As she reaches the trunk line, she stops the first worker she sees and waves her antennae round and round. The wounded worker then falls on the highway and dies.

Ants gather around to lick their dead nestmate. They taste her substances that are released instantly upon death. Then four ants carry her away. They take her the rest of the way home to their nest at the edge of the forest near the overgrown plantation along the Río Puerto Viejo. Some hours later, the corpse would be moved to the colony refuse chamber—a deep underground pit heaped high with the decomposing bodies of colony ants.

But the alarm goes out. After the message reaches the colony, it is passed through the dark corridors and tunnels. In seconds the soldiers begin filing out. Their long legs cover the distance to the foraging site in a few minutes, much faster than the workers. They arrive on the scene, mandibles gaping, in full threat posture, poised and ready. There has been a massacre. Body parts of leafcutter ant workers are everywhere. The pheromone smells from the alarm signals of so many ants hang in the air, mingling with the bitter smells of bala. The balas are nowhere to be seen. There is no one for the soldiers to fight. Seconds later a tiny ant called a "dacetine," a specialized solo hunter belonging to the Dacetine tribe, is waiting to snare a springtail when it happens to run afoul of the soldiers. It is dispatched in short order. Most of the leafcutter workers have survived, however, and are already back at work, snipping out leaf pieces and carrying them home.

Leafcutter ant soldiers can do little against the well-equipped bala, except perhaps through numbers, to keep them at bay. This is all that is usually necessary. The leafcutter soldiers are more than willing to put their bodies on the line. Yet, why had these balas attacked so fiercely—and apparently without even feeding on the leafcutter ants?

In the life of a colony, the death of individual leafcutter workers is of no real consequence; it is only a minor interruption in the colony's work for the day. With millions of workers, the loss of even several thousand has no impact on the survival of the colony. In a matter of days, the mother queen will lay enough eggs to replace them.

Next morning, the leafcutter scout goes out again. She passes the bushes she found yesterday, now neatly stripped. She has to walk around the bodies of hundreds of leafcutter workers, many now dismembered, being consumed by other species of ants, flies, beetles—scavengers all. The bala ants that killed them never even came back for the meat. Pausing only a few seconds, she pushes on, dedicated to her search for new plant material.

Minutes later, on one of her scouting zigzags, she comes to a massive crater, five feet across, at the base of a tree. All around are the huge droppings of an anteater. By smell, it is the site of a bala colony, now in ruins.

The massacre of the leafcutter ants the previous day is minor compared to the devastation of the bala nest. The broods and larval nurseries have all been turned out and destroyed. The anteater knew that the bala mother queen at the heart of the colony was the juiciest morsel of all. When it took the queen, it ended the reproductive potential of the colony. Although some colony members escaped with their lives, the anteater, in effect, eliminated the colony. The bullet ants that on the previous day attacked the leafcutter ants were some of these stragglers. Without their queen, their lives are meaningless. They might as well be dead. Their musky odor and the snapping of their jaws spelled alarm. But these alarm signals meant nothing to the leafcutter ants. They could not know why the big ants attacked them so fiercely. The balas were primed for battle—the last battle of the dead colony—and the leafcutter ants just happened to be in the way.

The scout knows none of this, of course and barely notices even the aftereffects. She has her job to do. Her life has meaning only in terms of the work she must accomplish for the colony. In her daily zigzag pattern, she scours the area for fresh new sources of leaves. Her only distractions are the smells that drift low over the earth, the strange smells that she remains ever alert for, the smells of so many predator insects, birds, and mammals. One of those smells is decidedly noxious, as she passes beneath a makeshift table set in the jungle. The human smell consists of an array of complex chemicals including human sweat, commercial antiperspirant, aftershave lotion, toothpaste, coffee, lingering red-hot salsa, laundry soap (clinging to clothes), and, above all, a fairly heavy dose of insect repellent spray. On both sides of the table are piles of detritus, leaf litter, pieces of logs, and the odd stray worker ant or a colony scurrying around. Two men sit at a table, one hunched over a microscope, the other banging a piece of wood into a white tray and then bending down to look through a jeweler's loupe. Myrmecologists at work. For a few seconds, as she scouts the vicinity, the ant's life would seem to be in danger. Would she be noticed? Would she be subjected to the intense curiosity of ant scientists? Today, at least, they have no interest in leafcutter ants.

Chapter Two

The Miracle Ant

These grand scenes . . . it is not possible to give an adequate idea of the higher feelings of wonder, astonishment, and devotion, which fill and elevate the mind.

—*Charles Darwin,*

DIARY ENTRY, APRIL 18, 1832, upon first seeing the tropical forest in Brazil, from *Journal of Researches into the Geology and Natural History of the Various Countries Visited by H.M.S. Beagle,* 1839

When you have seen one ant, one bird, one tree, you have not seen them all.

—*Edward O. Wilson,*

FROM AN INTERVIEW, *1992*

Hunting for Ants. Ed Wilson (at left) and Bill Brown search for unusual ants at La Selva, Costa Rica. One ant they meet is the little fire ant, Wasmannia auropunctata, *a fierce world conqueror.*

THE myrmecologists sitting hunched over a field table are Edward O. Wilson of Harvard University and William L. Brown, Jr., from Cornell University. Both have taken a break from spring teaching responsibilities to go to Costa Rica and indulge in their favorite pastime—ant treasure-hunting. On a hot morning, sun blinking through the rain forest canopy, they examine pieces of rotting wood, carefully breaking them apart above a white sixteen by twenty-inch plastic tray. They are intent on uncovering the middens of an ant colony. Identifying middens—the kitchen pantries that some ants keep—is a quick way of discovering the diet of an ant species. Often the pantry is more like a trash heap, with only the remains of food items. The "treasures" can include discarded beetle parts, legs from other ants, and even a few of the telltale headless termites.

"Now this is the way to do ants," says Wilson, beaming. His straight steel-gray and black hair, parted to one side, hangs down over his forehead almost to the top of his glasses as he focuses on the task. Tapping a chisel into a small chunk of decaying wood, he discovers a colony of ants, including larvae and pupae. As the wood slices open, it looks like a textbook cutaway. Moving their antennae around, the worker ants are yellowish orange against the splotchy white of the brood. The workers start running to and fro. They are trying to assess the threat. Wilson is careful not to breathe on the ants, but they are already sensing danger.

"What do you see, Ed?" asks Brown.

"Little fire ants. You know, *Wasmannia,*" says Wilson. "They're just minute. And, yes, we have some middens today."

Wilson leans down for a closer look with his one good eye. Wilson lost most of the vision in his right eye at age seven when, on summer vacation, he accidentally jerked a fish fin into it, causing a traumatic cataract. Months later he had to have the lens removed, leaving him with full sight in only one eye. The vision in his left eye, however, is sharper than sharp, 20/10, capable of reading a couple of lines below the usual level on an eye chart, capable of counting the teeth on an ant's mandible, capable of appreciating life on a minute scale.

"Beetle parts," he says. "Lots of 'em—the sternum and a detached head. Oh, there's the elytron [the wing] of the beetle and some remains, including the prothorax of another beetle. And an isopod [wood louse], some fly larvae, even some moth parts. Here, take a look, Bill."

Wilson passes the chunk of wood to Brown and inadvertently grazes a worker because he has no depth perception. The message—*Invader!*—soon gets around. The worker ants grab the comparatively large pupae and larvae in their five-toothed mandibles and carry them down several tunnels in the wood to temporary safety. Two other workers come out snapping their mandibles and waving their gasters, stingers poised.

The little fire ant, *Wasmannia auropunctata,* is a common rain forest ant that scavenges the forest floor, recruiting hordes of nestmates with odor trails and using group transport techniques to carry large and small prey home to the colony. *Wasmannia* is named after the German Jesuit anti-Darwinist scholar Erich Wasmann who spent much of his life in the late nineteenth and early twentieth centuries researching ants and, bucking Darwin, saw in them manifestations of divine powers. Wasmann made many solid contributions to entomology, and getting an obnoxious ant named after him was presumably part of his reward.

They're called "little fire ants" because of their size and fierce sting. They are much smaller than the notorious imported fire ants of the southern United States, but their sharp sting is just as painful. They are also difficult to eradicate —just like the larger fire ants.

Brown places the piece of wood containing the colony under his portable high-power microscope. Right away he notices something oddly familiar in the midden's scrap heap. "They have another ant here, too," he says. "It's our old friend *Gnamptogenys*. Unfortunately, it's just the head; I guess they've eaten the rest. Maybe they picked him up along with the beetles."

Brown worked out the taxonomy of *Gnamptogenys*, a large worldwide genus in the subfamily of ants called Ponerinae, which Brown has focused on over the years. The ponerines are the solitary hunters of the ant world, though many species of other subfamilies also hunt. Solitary hunting tends to be the primitive condition in ants. Some ants have evolved away from the original solitary hunting habit; others have retained it. Some have come to specialize in particular prey; others have taken to hunting in groups. Still others, such as many species in the more advanced subfamilies called Myrmicinae and Formicinae, have shifted their diets and behaviors toward eating everything or specializing in plants or mushrooms. In this case, the little fire ants, fierce myrmicine ants, have captured and killed a *Gnamptogenys* huntress.

"What else, Bill?" says Wilson, making a list.

"All kinds of live mites running around—did you see those?"

"It's not surprising," says Wilson. "But no sign they're eating mites?"

"Nothing," says Brown. "They're sure messy eaters. Oh, here's another large beetle part."

"Well, we know they're scavengers," says Wilson.

"Yes, general insectivores," says Brown. "And apparently big on beetles. The beetle fragments last longer than other things, of course. They're certainly not mite specialists—or any other sort of specialist."

"All these tiny mites; there have to be some ants that specialize in them," says Wilson.

"Can you see the big queen?" he asks. "I'd love to take a colony back to the lab sometime. No one's done an in-depth behavioral study of the little fire ant."

"No, they must have carried her off, or she's buried here somewhere," says Brown. "Better watch it with these little fire ants. They'll get you."

"Well, we're off to a roaring start," says Wilson. "Okay, let's find some more colonies. I'll show you where I foraged last year—the best piece of forest for ants you've ever seen."

"I know it's good here, Ed," Brown says. "I've described more than one ant from Costa Rica."

Since 1985, Wilson has been visiting La Selva every year or two. Brown was last here in 1973, and before that 1966. Both have been involved with the Organization for Tropical Studies (OTS), a consortium of U.S. and Costa Rican universities, including Harvard and Cornell; the OTS has used this reserve and biological station as its primary center for tropical research since 1968. The name "La Selva" means simply "the forest" in Spanish, and this basic name is appropriate for the elemental role this forest has played in terms of research and conservation in the tropics. It's an ideal locale for study focusing on ecology or individual species. Besides the diversity of ants, there is a rich flora from primitive cycads to many species of orchids and canopy trees and a fauna that includes many frogs, snakes, birds, several species of monkeys, anteaters, sloths, and thousands of insect species. There are experimental plots for study as well as a strictly protected reserve, all in a classically beautiful lowland tropical rain forest setting.

Rising from the makeshift table, Wilson and Brown fold up their portable chairs and gear, and they're off down the path. The leafcutter scout, meantime, has forged ahead on its own morning mission, neatly eluding the two men. Both about six feet tall, they look like two monks striding methodically with head and shoulders bowed reverently in the classic myrmecologist's stoop. Poor posture seems the inevitable consequence of a lifetime spent combing the ground and lower foliage for ants. Also, their eyesight is not getting any better. "Either that or the ants are a damn sight smaller these days," says Brown, clearly hunching over more than Wilson. They are two prospectors looking for the big strike—both dressed in faded browns with light tan gabardine trousers tucked tightly into old hiking boots that have been sprayed with bug repellent and dusted with sulfur powder to discourage chiggers and other unwanted insects from attack. Both are nothing if not absorbed and yet are surprisingly conversational, although Brown's incessant mumbling of having forgotten his machete is beginning to sound like a broken refrain. As amusing as the idea of hunting ants with a machete is to Wilson, Brown feels naked and helpless without it. Part of it is a difference in style: Brown likes to excavate ant colonies in large logs, arguably best split by a machete or axe; Wilson is a twig specialist for whom a penknife is sufficient.

As they trade one-liners, barbs, and fish-style ant stories, Wilson and Brown march deeper into the forest, and the particular holy grail of this expedition emerges—to find "the miracle ant." It is a shiny black ant thought to live in small colonies. Wilson and Brown have seen one or two alive before. In the few sightings at La Selva, lone workers were seen foraging. But no one has yet found a queen or colony or studied the ant's natural history. The big question revolves around the large, pitchforklike mandibles. When the mandibles are closed, the long, slender teeth extend even past the back of the head. "Imagine discovering a human civilization in which all the adults constantly wear helmets adorned with sabers," says Wilson. The shape of an ant's mandibles usually gives a clue about what it eats and/or how it defends itself. But the miracle ant has the most unwieldy set of jaws imaginable. What does it use them for? Do they have something to do with food specialization? And how does the miracle ant handle her young? "Delicately," says Wilson with a smile. In any case, that's how the species earned its scientific name—*Thaumatomyrmex.* In Greek, *thaumato* means miracle or marvel, and *myrmex,* ant—the miracle ant. If Wilson and Brown find a worker, they will try to follow her back to the nest, then study the colony in its natural habitat.

Wilson came up with this quest. The search for the miracle ant is not the sole purpose of the expedition; it is more of a metaphor for the kind of major find that Wilson and Brown seek. Longtime collectors have to search out the rare ants and unusual ant behavior. Together, Wilson and Brown have collected thousands of ant specimens and sizable portions of both the Cornell and Harvard collections. But their days of collecting all ants are long over. They have to go for gold. And defining a search for some species makes a big find more likely to happen. Wilson jokes about erecting "wanted posters" for the miracle ant, and he later sketches a few mug shots and passes them around the station. The prize for finding one alive is a noisy acknowledgment in the paper describing the ant, maybe even coauthorship, and certainly a case of beer. "If you find it," he tells a science journalist, "I'll write an article about *you.*"

Besides the search for miracle ants, Wilson and Brown are out here in the real world to try to put life into perspective. It is their way to renew the spirit, to make an annual pilgrimage and "worship," as Wilson puts it, in the great natural cathedral of the tropical rain forest, where life has its most diverse expression. For both of them there is still the basic boyhood thrill of prospecting that goes back to Sir Arthur Conan Doyle's *Lost World,* one of Wilson's favorite books growing up. Both have found a few lost worlds in their day, at least of the Lilliputian variety, and know that amazing things turn up when you are least looking for them. But to allow for that possibility, you have to get into the field.

It has been thirty years since Wilson, sixtyish, and Brown, in his mid-sixties, have gone collecting together. They have collaborated on papers and visited each other often over the years, but this trip to La Selva is something of a reunion in the field for two friends and colleagues. In the interim, Brown has visited almost every country that has tropical forest, as well as many remote islands. Brown has seen or collected an estimated four thousand ant species worldwide, more than any other entomologist. Wilson, who has close to two thousand ant species to his credit, was beating Brown based on his intensive collecting in the 1950s, but since then Brown has surpassed everyone. Wilson calls Brown "probably the most widely traveled and productive ant collector of all time." That "probably" is no more than scientific convention. Brown's warm stories about western China, India, Java, New Caledonia, Madagascar, Mauritius, and the Mato Grosso of Brazil all seem as if they happened yesterday—they sound that fresh. Many are accounts of the ant that got away, and every story ends with a sumptuous native meal and the ritual imbibing of local beer ordered in local dialect. Brown could lecture long and brilliantly on the best local ales in remote outposts. He knows them all—outposts and ales—and remembers them almost as well as the names of several thousand ant species that are on the tip of his tongue. Wilson, despite intense practice to learn Portuguese before going to the Amazon in the early 1980s, is unilingual, and his adventures in New Guinea, Fiji, and in the Latin American tropics are usually with colleagues or guides who can speak the language and smooth the way. Wilson still retains some boyhood shyness. Brown, on the other hand, slips into the local rhythms and superimposes his own beat. He is a story waiting to happen. There was that time in the Dominican Republic that he and two younger collectors had a car "accident" on the way to the airport—they were rammed on purpose by local desperados. It was either invite the police to help them and miss their plane and probably worse, or produce a fat bribe to be on their way. Brown's Spanish and coolness under pressure managed to get them off. Brown's stories often come without apparent context, and Wilson has to stop him to ask how he managed to jump from stalking new species of ants in Mauritius when a rainstorm nearly blew him off a mountain to getting questioned for murder simply because he looked suspicious, hacking the ground with a machete beside some highway in another hemisphere and in another decade. Wilson tells stories to illustrate specific points. He speaks as logically as he writes, adapting the material for any audience and parenthetically defining terms that he senses the listener isn't grasping. But both men can draw on their experiences with many species and ecological situations to try to elucidate an idea. Essentially, however, it remains the world from two ant fanatics' perspective.

"This place is loaded with dacetines," says Wilson, stopping at a rotten log. "This is the third one I've seen in five minutes. We must be getting into prime dacetine country."

"And look at this *Strumigenys* colony," says Brown, referring to a long-jawed dacetine species at the other end of the log. "Pardon me while I have a good drool."

Ants from the Dacetini tribe, or dacetines, are part of the Myrmicine subfamily found around the world, but the New World tropics is a stronghold. In Costa Rica and Panama alone, there are some thirty-two species—most of them likely found at La Selva. Both Brown and Wilson cut their teeth on dacetines, which were once considered rare by ant collectors, and these elegant, specialized predators taught the two young myrmecologists many things, including how to work out the evolution of an ant tribe.

Dacetines are "ornately sculptured little things," as Brown puts it. To myrmecologists, the slender, delicate torsos, long mandibles, and finely striated abdomens make them "really attractive ants." They are decorative; some species have white or yellow spongy collars around their thin waists, and their body hairs look like clubs, scales, or even "sinuous whips."

Wilson reaches down and, using tiny, flexible forceps, grabs the little dacetine gently by the abdomen and holds it up to his loupe to admire it. Though he can see it well with his naked eye, the hand lens lets him enter the ant's world in an enhanced, complete way.

"She's caught something," says Wilson. More than superficial beauties, dacetines are, in Brown's words, "intriguing little devils." Clutched in her long mandibles is a collembolan, or springtail, a soft-bodied, drably colored arthropod, an insect relative, about three-sixteenths of an inch long. The specialist huntress ant has just landed its tiny, primitive prey—the final act in the classic predator-prey drama, but at a scale rarely considered outside entomology. And springtails are hard to catch; they can't fly, though they jump like miniature kangaroos, as Wilson puts it. They have a fork-shaped appendage that is doubled up beneath the body and acts as an unpredictable spring. Most of the time they probe through the leaf litter, and eat decaying plants and nematode worms, but if they are disturbed, they spring away. Springtails are numerous and ancient. Of all known insects, the springtails, with some two thousand species in a worldwide group, are the oldest and have flourished almost unchanged for 400 million years, since the Devonian period. But for the springtail, the ants have evolved the better springtrap.

To catch a springtail, dacetines have to hunt with slow, deliberate movements. As the ant picks up the odor on her antennae, she stops in her tracks and opens her mandibles wide—to as much as 180 degrees. The mandibles are locked into place—the trap is set. Then the ant projects sensitive hairs above her mouth. Using these "hair trigger" probes, the ant inches toward the springtail. When the hairs make contact, the ant lets her mandibles snap shut, impaling the springtail on the lower mandibular teeth.

Brown, fascinated with the springtrap jaw, had actually taken it apart to see how it worked. Inside he found a tiny J-shaped tendon, hard and tense as a spring with the muscles attached to a cylinder that rotates until the jaws snap shut.

Both Wilson and Brown have a certain admiration for the solitary specialist hunter—the ant equivalent of the jaguar or mountain lion. Maybe they identify. Finding food and then laying recruitment trails to summon a good portion of the colony to help is impressive as a form of social organization and communication, but it's a fairly common sight in the ant world. The rarer story of the solitary hunter bringing back the specialized catch to the colony—as in the dacetines—attracts the imagination and the prospecting sense of myrmecologists. They are, after all, to a large extent, solitary specialists themselves.

For a few seconds Wilson holds up the *Strumigenys* for both of them to admire. A hundred shared stories come to mind, but both men are quiet, reflective. Wilson sets the little *Strumigenys* down on the ground and it slips away.

WHEN BROWN AND Wilson first met in Cambridge, Massachusetts, in 1950, Brown suggested that Wilson do the definitive survey of the semitropical dacetines of the deep South. At the time, Brown had recently arrived from Pennsylvania State University and his home city of Philadelphia to work on his Ph.D. at Harvard. Southern-born and -bred, Wilson, then twenty-one years old, found Brown to be a slender, nervous, chain-smoking young man, while Brown, then twenty-seven, saw Wilson as an earnest, flat-footed, gangling lad. Wilson was delighted when Brown treated him as an equal. What most impressed Wilson about Brown was that he was a complete and utter ant fanatic. So was Wilson. Then a junior at the University of Alabama, Wilson had field experience with the dacetines and imported fire ants from the backwoods of Alabama and Florida where he had spent most of his time up till then, but he didn't know that dacetines were anything special. His life ambition before college had been to ride around in one of the green pickup trucks used by the U.S. Department of Agriculture's extension service as a government entomologist, advising farmers on how to control their insects. Even before he left the University of Alabama he realized that goal as an entomologist with the Alabama Department of Conservation. But the University of Alabama gave him the idea that people could do pure science and get paid for it. And Brown, who was studying dacetines throughout the world, gave Wilson his first taste of the big world of ant research and ants. One of the fascinating aspects of dacetines, Wilson soon learned, was that each species has a distinctive way of capturing its prey. And when you start comparing them to others around the world, it gets really interesting.

Usually, however, myrmecologists don't find the ant with the prey in its mouth. They check the colony for middens. For the many ants that have no middens there is another alternative: an invitation to Wilson's Cafeteria. Wilson devised this technique, now used by other entomologists, when he was sorting out the dacetines. To stage a cafeteria experiment, Wilson sets out various possible prey around the ant's nest and then watches what happens. Posing as a sort of busboy cleaning up after the ants, Wilson carefully notes exactly what kinds of food are taken and which are actually eaten.

A key early paper that Wilson had published in 1953 explored the comparative ethology of the Dacetini—before he really even knew what ethology was. Decades before their shared Nobel Prize, Konrad Lorenz, Karl von Frisch, and Niko Tinbergen were pioneering a new approach to the study of animal behavior, called ethology. Before ethology came along in the 1930s, animal behavior studies had been driven down a blind alley of manipulating behavior in laboratories to try to learn about it. Observations in the wild were thought to be old-fashioned—suitable for boys and clergymen with time on their hands. Ethology—which was a return to some of Charles Darwin's original ideas with an updated approach—made the observation of wild animal behavior into a strict science, studying the animal in the context of its life cycle, ecology, and evolution. The goal of ethology is simply to understand natural behavior such as courtship, social signals, nesting, and territorial behavior. To do this, many methods, including experimental techniques, can be used. Wilson heard Lorenz speak at Harvard in late 1953, and he describes it as a "thunderbolt." There was the shock of recognition, a tacit approval for the kind of approach that he and Brown were already taking with ants, and there was a host of new ideas that could be applied to myrmecology. Soon after, Brown and Wilson published their worldwide synthesis of dacetine biology. They traced the evolution of food habits and correlated that with social organization and biology. Through this foray into "socioecology," as Wilson calls it, they found that the most primitive-looking dacetines—such as one awkward beauty with extremely long mandibles called *Acanthognathus teleductus,* which Brown first found in the Colombian Chocó and which Wilson recently saw at La Selva—have habits that include foraging above ground for various insect prey. But as the dacetines evolved, they became more and more specialized in springtails. At the same time they began burrowing into the soil and wood, and smaller individuals gained an advantage. Over many generations the ants started spending more and more time underground, moving their nests to these safer spots. As their prey got smaller, their mandibles tended to get smaller, too. The key anatomical distinction of *Acanthognathus,* however is the eleven antennal segments, shared only with the *Daceton* species. These are the most of any of the several dozen dacetine species and are considered the primitive condition. Brown does not recognize *Daceton* as a true dacetine, however, so in his view *Acanthognathus* is the most primitive.

Published in a widely read biological journal, the *Quarterly Review of Biology,* Brown and Wilson's landmark paper on the dacetines received scant attention. Before their work, individual studies of the biology, ecology, and food habits of ants and other species had been reported, but no scientist had done such a comprehensive evolutionary study. Wilson says that it was one of the first sociobiological studies in that it emphasized the adaptation of social traits. But, according to Brown, in the 1950s people really weren't ready for the idea of adaptiveness of social systems, not to mention of the ants. "When primate researchers started looking at gorillas and chimps the same way in the 1960s," adds Wilson, "it was hailed as a novel approach. People are just more interested in primates. But Bill and I had really worked it out in more detail than the primate people did—or even could—because you can work so much faster with ants." To this day Brown and Wilson's findings have had little impact and remain obscure except to certain entomologists and a few others who grasp the importance of social insects.

In fact, if an evolutionary biologist were given three years to test a set of general evolutionary hypotheses or to answer questions about social behavior, with extra points awarded for new insights, the best choice would not be a bird or mammal species—even though these are much more common research subjects. There might be no wiser choice than ants. The reason is that most insects have shorter life cycles suitable to research. Also, one can do experimental research with insects. Ants, in particular, are extremely numerous, and worker ants, taken for research, can be quickly replaced by the colony. As useful as fruit flies are to genetic research, ants may be the ideal animal to learn about social behavior and many related aspects of evolution.

Wilson and Brown both say that the ants have given them everything. They mean that literally. Both have seen the world, risen to prominence as teachers and scholars, and achieved a rich intellectual life thanks to ants. Ants gave Brown a deep understanding of how evolution by natural selection works and an appreciation of the importance of learning everything possible about a group of organisms before attempting to classify it. Through ants, Wilson achieved his understanding of social behavior, which led eventually to sociobiology. His knowledge of how ants are distributed on islands was the key to his work in island biogeography, the science that deals with the distribution of animals and plants around the world, as well as his recent biodiversity studies.

Wilson cites several turning points on the road to ants. The first came from his physical impairments. After his fish fin accident, his one good eye limited his ability to observe birds, mammals, and reptiles. He also had a slight hearing loss, especially in the upper registers, which meant that bird song was lost to him. But his especially good vision in one eye was ideal for insects. The second turning point was a World War II shortage of insect mounting pins, then made in Czechoslovakia. At age fifteen, when Wilson felt he had to choose an insect group to concentrate on, he first thought of flies, but they required the hard-to-get steel pins. Wilson turned to his second choice, ants, which could be collected in simple medicine bottles filled with ordinary rubbing alcohol and preserved indefinitely.

The third turning point was Brown. It was Brown who persuaded Wilson to go to Harvard, and Wilson says Brown has been the single greatest influence on his scientific life. Brown first urged Wilson to "work on ants worldwide" —a debt that Wilson has never forgotten. In 1992 when Brown retired from teaching (though not from ant work), Wilson thanked Brown for saving him from *The Ants of Alabama,* his first scientific ambition. Brown realized early on that myrmecologists, as much as evolutionary biologists, had to take the worldview. It doesn't matter which group of animals or plants you work on, but it is essential to learn to identify and compare related species from around the world in order to assemble evolving lines and show the different stages of advancement from the primitive to the highly evolved. If one examines the greatest number of traits derived from primitive species, the evolutionary history or family tree of a group of animals, the phylogeny, can be determined. This is the often elusive goal of the modern taxonomist. Even today Brown spends much of his time on monographs, whether of a single genus or an entire tribe, sinking or discarding species designated by local entomologists. (To taxonomists, throwing away names is referred to as "sinking" a species.)

Brown says, "They just don't have a clue that their special ant is found in other parts of the world. Most of these hot finds are closely related to, or even the same as, species already named." Brown, who dislikes patronyms—species named in a person's honor—even eliminated an ant species bearing his own name.

"You sank your own species?" says Wilson when he hears.

"I didn't mind at all," replies Brown. "It wasn't a real species. I was glad to get rid of an extra name."

Brown takes few things as seriously as the awarding and naming of a new species. Instead of patronyms he would rather create a name based, first, on its evolutionary history and, second, on a characteristic of the species—as an aid to memory. But there is a dacetine ant called *Strumigenys wilsoni.* Brown says with a sudden smile that he simply could not find much distinguishing it. Or so he tells his friend Wilson. And there is another dacetine—a rare ant genus from Cuba named after Brown's wife Doris: *Dorisidris nitens.* Both are elegant, handsome dacetine ants, so it could be argued that the names are fitting. For Brown, the acknowledged living master of ant taxonomy, to name an ant for you is indeed a rare, high tribute.

With their collaborations on the dacetine work, Brown and Wilson began to lead the way in the new systematics of ants, following the species concept of Harvard evolutionary biologist and ornithologist Ernst Mayr. Darwin had said essentially that a species was what an expert said it was, but Mayr added that it had to be defined on a strictly biological basis—every individual of a species had to be able to mate with each other. In his 1942 book *Systematics and the Origin of Species from the Viewpoint of a Zoologist,* Mayr presented a strong case for the biological species concept, connecting the process of species formation to genetics and thereby opening up natural history to scientific analysis. He effectively called young evolutionary biologists to an exciting quest. Advancing this approach, Brown and Wilson and a handful of other myrmecologists began correcting the "descriptive atrocities" of early myrmecologists around the world, many of them "feckless dabblers," in Brown's words, or those who published too fast and furiously, or engaged in "description for description's sake" and were thus guilty of "taxonomic irresponsibility." In an effort to sort out the mess, they revised one genus after another in a series of rigorous monographs that for the first time began to tell the true history of the world's ants and how they had radiated throughout the world, taking over niches and evolving wherever they went. Brown and Wilson used behavior and ecology, as well as physical appearance, as taxonomic characters. And they did it based substantially on what they found and witnessed in the field. Their findings are part of the exciting early chapters in the story of the ants, which both men are still engaged in unraveling to this day.

By 1951, Brown had graduated from Harvard and departed for Australia and points east on one of the first Fulbright fellowships. In late 1954 and early 1955, Wilson traveled for ten months around New Guinea and the South Pacific, sponsored by Harvard's Society of Fellows and the Museum of Comparative Zoology (MCZ). For both men these were opportunities to study rare ant faunas in their natural habitats. The trips were also essential *wanderjahren* that provided a world perspective as well as an appreciation for the astonishing diversity of all life. After Australia, Brown returned to Harvard in 1952, then moved on to Cornell University in 1960, becoming professor of entomology, while remaining an associate in entomology at Harvard's Museum of Comparative Zoology, where he continued to do much of his work. Wilson stayed at Harvard and worked his way up to full professor and curator in entomology at the Museum of Comparative Zoology. The MCZ ant collection, first built up by William M. Wheeler and substantially enhanced by Brown and Wilson, is the finest ant collection in the world—about a million specimens representing some forty-five hundred species. This is about half of the known world total of named ant species.

Yet when it comes to ants, Wilson and Brown seem to have a casual attitude, as if they have seen it all. On a typical collecting trip they pass dozens of species per hour that they seem to ignore. In fact, they ignore nothing. Some species they can identify without even bending down for a closer look, just by the way the ant scurries across a leaf or lopes along the ground. Is the ant shy or bold? Slow or fast? Solitary or in formation? Does it have long jaws or short? Hairs or not? Each ant species has its own *gestalt,* as Wilson puts it. The long processions of leafcutter ants, with each ant carrying a piece of green leaf or flower, are recognized at a distance of one hundred feet or more. Other ants must be held up to a magnifier, from 5x to 8x or greater, to see some minuscule key feature that is diagnostic. At up to an inch long, a bala ant is more than twice the size of most ants, but it is also as different in overall appearance and behavior from a leafcutter ant or a dacetine as a mountain gorilla differs from a chimpanzee.

And Wilson and Brown have emphatically *not* seen them all. Even using their rigorous standards, it is still possible for a myrmecologist to discover a new species or two or even ten on every visit to the tropical rain forest. When a new species is suspected, Brown and Wilson start getting quietly excited, although Brown sometimes turns cross if the species should have been included in a new or soon-to-be-published revision of his. Of course, then it must be carefully compared to other ants from the collections at Harvard and Cornell and around the world—the U.S. National Museum in Washington, D.C., the Natural History Museum in Britain, the Los Angeles County Museum of Natural History, and the Australian National Insect Collection, among others. The route to new speciesdom can be instant, but it is more often long, winding, and uncertain.

DAY AFTER DAY, as Brown and Wilson penetrate deeper into La Selva in a week of solid prospecting, they turn up many more insect species to satisfy their curiosity. They meet an earwig tenderly guarding her egg pile, displaying the intense parental care that earwigs are noted for. In an abandoned termite nest they find stingless bees that have no stings but do get in one's hair and, when they bite, inject mandibular gland secretions that can irritate. They run across the excavated bala, or bullet ant, nest, destroyed by anteaters. Several bala—among those that had attacked the leafcutters—are still milling around as if waiting for their mother queen to return. "Maybe there are some larvae left in one of the chambers," says Wilson. "Even though the queen is gone, the colony may be able to survive if the larvae has a new queen and can somehow get out into the world and into the next generation."

But the anteaters have eaten the juicy larvae. A little later Wilson is turning out a cacao pod looking for ants and starts poking with his finger at some white objects inside. Suddenly he drops the cacao pod, shudders, and steps back. The white objects are tiny eggs, and an enormous wolf spider, camouflaged until this instant, is jealously guarding them.

"I just don't like spiders," Wilson says half to himself, half to Brown, who well knows Wilson's antipathy to arachnids. While spiders are unwelcome, neither Wilson nor Brown minds being ant-bitten in the cause of science—with the exception of the bala and a few other large ponerines. Many ants will bite or sting given a chance, but usually the wounds are not painful, certainly less so than an ordinary bee sting. The myrmecologist at work dismisses most ant attentions, considering them more or less like the playful nips of a puppy. At one point Brown has a ponerine ant called *Hypoponera* on his arm and is trying to steer the animal into a collecting bottle. As Brown searches for the forceps, he tries to keep one eye on the ant held between thumb and forefinger.

"He's going to sting me," says Brown matter-of-factly, with no sense of alarm. "He's stinging me. Oh, I've been stung."

It's all in a day's work.

Seconds later Wilson knocks a huge bullet ant off the back of Brown's shirt. "You don't want that nudging up to you, Bill."

"Oh, I guess not, Ed," says Brown, watching the big ant right itself on the ground and strut off.

In fact, balas are not only the chief ant to beware of, but they are the most avoided insect in the Neotropical forest. They provide a topic for constant conversation among researchers young and old, especially over an evening beer. One neophyte researcher, a woman, had been stung on the foot as she slept in the generally insect-proof bunkhouse. Well before dawn, her screams roused her fellow researchers as well as half the jungle. A male researcher using outdoor sanitation facilities "with a great view" sat on a bullet ant and immediately shot up, yelling as he shot out of the privy; his pants around his ankles tripped him, and he fell flat on his face. The sting is variously described as "sharp," "nasty," but more often "paralyzing." Tropical entomologist and author Adrian Forsyth once earned four bullet-ant wounds on his upper back. He called it the most painful, non-lethal encounter one can have with any animal in the tropical forest. It is certainly the biggest day-to-day hazard at La Selva. Certain snakes, such as fer-de-lances and bushmasters, can kill, of course, and at least one researcher a few years ago died from a fer-de-lance at La Selva. But a bullet ant can and does much more commonly ruin a day, and the soreness lasts for a week. One sting from the bullet ant is sufficient to dampen many a researcher's enthusiasm about ants, insect diversity, and the tropical rain forest—in one blow.

"They do have a godawful sting," says Brown. "Once I got three or four stings in a row. Gosh, I had to sit down for a few minutes and catch my breath." More like thirty seconds, given Brown's enthusiasm.

As they reflect on bullet ants, spiders, and other minor inconveniences of the tropical rain forest, the ant species count climbs ever higher. They find perhaps ten species the first evening after arrival, then about forty the first full day, a total of more than one hundred by the end of day two. And so on, building to the big finds of day three. Some of these were repeats, but many were not. A characteristic aspect of the tropics is the species diversity—the astonishing fact that whether trees, butterflies, bees, or ants, the distribution of individual species is patchy, and for days of intensive collecting covering a square mile or two, one can keep coming up with new species until the diversity levels off. By contrast, temperate forests tend to have many more individuals of the same species living close together.

The diversity of all plant and animal life follows a common pattern. The diversity is highest in the moist tropics. The number of species declines as conditions become drier and especially colder, moving farther north and south into the temperate zone toward the poles. Costa Rica, which is the size of West Virginia, has rain forests that are home to 820 species of birds, more than in all of North America. Many more kinds of trees and other plants are found in Costa Rica than in similar-sized temperate and polar regions. Much larger Greenland, for example, has only 56 species of breeding birds.

The diversity of ants follows suit. On a single tree in the Tambopata Reserve of Peru, Wilson identified 43 species of ants belonging to 26 genera—about equal to the entire ant fauna of the British Isles. Brown, assisted by his wife Doris, once recovered 81 ant species from a single log near Manaus in the Brazilian Amazon. It took only a day and a half, thanks in no small part to Brown's machete. Some of these were new species. And in one eighteen-acre plot of rain forest near Puerto Maldonado, Peru, Wilson and two fanatic younger collectors, Stefan Cover and John Tobin, set out to break all records for one locality. They did it, collecting some 275 ant species. The ant diversity at La Selva is even more spectacular. Resident La Selva myrmecologist John H. Longino once collected 60 ant species from the crown of a single *Carapa guianensis* tree, a common rain forest canopy tree of the mahogany family that grows up to 150 feet tall. In all of La Selva, Longino knows of about 350 species of ants. The ant diversity falls off dramatically in the cooler temperate zone, and only three known ant species range north beyond the tree line of the Arctic. In the whole world, approximately 9,500 ant species have been described, belonging to some 300 genera. There may be 15,000 total ant species, according to some myrmecologists, 5,000 to 6,000 still to be discovered, mostly in the tropics. Wilson, however, thinks the total may well reach 20,000 to 30,000 species. In any case, says Wilson, "in less than a few acres at La Selva, there are enough different kinds of ants to occupy the lifetimes of a dozen scientists."

Almost any sizable sample of ants from the tropical rain forest is likely to have new species of ants. With so many yet to be discovered and named, there are plenty of trophies and fascinating stories left for young aspiring collectors. "Finding new species is no trick," says Brown. "But figuring out how it relates to existing species is the challenge. Young myrmecologists today tend to stick to the more obvious species—leafcutter or army ants—because these are simple to identify. But a lot of things about social behavior, feeding habits, worker or queen behavior, and so forth come to light only by examining closely related species." A taxonomic understanding of whatever one is working on is essential. In many fields of biology, more courses in taxonomy—and many more taxonomists—are needed. A lot of biological work depends on taxonomy, but that does not mean that taxonomy is a lesser science. Brown once overheard a noted ecologist at a conference calling taxonomy "the handmaiden of ecology." Brown, incensed, stood up and said, "No, a handmaiden is a girl who works in a massage parlor." To Brown, an absurd statement required a rude retort.

Brown is a serious taxonomist who belongs to an even smaller group of serious systematists. In some quarters and contexts, the terms are used almost interchangeably, but not all taxonomists are systematists. Most taxonomists, typically associated with national museums, are responsible for so many species that often there is only time for classification. Systematists—such as Brown and Wilson, who are engaged in the wider adventure of evolutionary biology—are experts on a group of species. Systematists are primarily interested in diversity, including classification, but range freely into other aspects of the biology of their favored group.

Breaking for lunch, Brown hears about a plan to catalogue the insect diversity of La Selva and shakes his head.

"What a waste of effort," he says as they sit down to beans, rice, and a main course surprise. "These local and regional faunal studies are next to useless."

Wilson comes back sharply: "No, Bill, I disagree. There's not always enough time to do the kind of careful, painstaking monographs you've done for the ants worldwide."

"But it's all going to have to be redone," says Brown, shaking his head. Brown has spent decades sifting through the accumulated mess of what he calls the "garbage heap" of taxonomy. He has disparaging words for would-be taxonomists of every era who have made his life miserable from time to time.

He stirs his fork into the rice and beans, mixing them up, as if to illustrate how messy things can get.

"Honestly, Bill, I think we have to support these efforts," says Wilson. "We're on a sinking ship. We're losing species every day."

But Brown shakes his head and pours on more salsa. "Beans 'n' rice. Rice 'n' beans. By the end of the week," he says, clearly fuming, "we'll be using a bottle of sauce every meal." He mutters something offhand that sounds like "So much for the diversity of the rain forest!"

Then he sees a white ant walking across the table. "Look, Ed!" And Wilson says, "Here's a different species crawling up the table leg over here." By the end of the meal, Wilson and Brown have found three species of ants living in the La Selva dining hall alone. They are back in business.

But the skirmish between Brown and Wilson, which flares again after lunch and off and on all through the week, is something like the difference between the pure scientist and the pragmatic scientist who is also a conservationist. Both Brown and Wilson are committed to conservation. Brown, however, is a conservative purist when it comes to ant research, and in a perfect world he's right. But Wilson knows that the world can't wait for all the careful taxonomy of one or even a few Bill Browns, no matter how dedicated. Even ten Bill Browns would not be enough to get the ants all sorted out before their habitats, in many parts of the world, are disturbed or destroyed.

BROWN TENDS TO get fussy when it comes to myrmecology and the taxonomy, or systematics, of ants, but he clearly loves the tropical forest and knows the names and stories of plants as well as many insects and animals. At his home north of Ithaca, New York, he raises hundreds of species of orchids and hangs them from the oak trees around his house, re-creating some of the lush, epiphytic feeling of the rain forest. In the 1960s, Brown in effect defended biodiversity with his support of Rachel Carson's work. One of Brown's papers formed part of the source material for *Silent Spring.* Brown was asked to review the book. After *Silent Spring* hit the news in 1962, sounding the alarm about the danger of DDT and other pesticides, one entomologist sympathetic to pesticide control asked Brown if he didn't think Carson's statements on insecticide use were extreme. Another "applied" entomologist sent Brown a questionnaire asking if Carson had accurately represented his views. Later, during hospitality suites at entomology conferences, Brown defended Carson against disparaging remarks from pesticide company representatives. "We entomologists have to stick together," they intimated. "Pesticide salesmen," Brown says, shaking his head. "The errors were minor, and what she said basically was and is true." Brown reviewed *Silent Spring* favorably and later visited with Carson before she died of cancer in 1964.

Like Brown, Wilson was concerned about the loss of biodiversity early on, and his eclectic nature and love of the Big Biological Picture led him to collaborate with several theoretical ecologists. The first was Robert H. MacArthur, a Canadian who was then professor at the University of Pennsylvania and later at Princeton, and a specialist in biogeography. Wilson wondered why the ants he found on the South Pacific islands seemed to have certain patterns of diversity and distribution. There are more species on large islands than on small islands, and when Wilson looked carefully, he saw that with every tenfold increase in area, the number of species doubled. He saw, too, that as ant species spread from Australia and Asia to the smaller islands between, they eliminated species settled there earlier. What did it all mean?

Adding the ant data to MacArthur's work on birds, the pair came up with the first theory of species equilibrium. It turned out that the rate of immigration to an island would increase until the island reached a dynamic equilibrium. From then on, the number of species would stay the same. In general, once equilibrium had been reached, when a new species immigrated to an island, an existing one would become extinct. Wilson and MacArthur's theory also predicted that the extinction rate would be faster on smaller islands since the populations are smaller and more liable to extinction. After an invasion, however, smaller islands could return to equilibrium faster.

They tested their theory on Krakatoa, a small island between Java and Sumatra in the Sunda Island group of the East Indies. A volcanic eruption in 1883 had obliterated all life, and between 1884 and 1936, scientists mounted expeditions to monitor the recolonization of the dead island, particularly the land and freshwater bird life. The first plants and animals arrived within a year, and after thirty-five years a dense forest covered most of the island. For an island of that size in the Sunda group, the MacArthur-Wilson model predicted equilibrium at thirty bird species. In fact, after thirty-eight years, equilibrium was achieved at twenty-seven species. During the next fourteen years, five species arrived and five disappeared. The facts closely fit the theory.

MacArthur and Wilson's 1963 paper in *Evolution* and a monograph in 1967 launched the study of island biogeography. It had been Conan Doyle's *Lost World*—set on a continental plateau, a sort of unexplored island—that made Wilson a compulsive island lover and set him on the path that led to his codevelopment of the theory of island biogeography. But in 1972 when the science was still very young, MacArthur died of cancer. Only forty-two, he was then one of the world's leading theoretical ecologists, and his work still has an impact today. The theory of island biogeography, modified and expanded, now applies not only to oceanic islands but to "habitat islands" such as tropical forest parks and nature reserves surrounded by slash-and-burn farms and cattle ranches. In 1983, more than a decade after his work with MacArthur, Wilson visited the species-rich Amazon to collect ants and see firsthand the Minimum Critical Size of Ecosystems Project. The latter has been a bold long-term test of MacArthur and Wilson's early work. Researchers from the World Wildlife Fund–U.S. and Brazil's National Amazonian Research Institute have been trying to learn how best to design tropical reserves in order to save as many species as possible in pieces of rain forest that can be set aside. By 1989, more than half of the world's tropical rain forests had been cut or seriously degraded. Nearly 55,000 square miles are still being destroyed annually, according to estimates by British conservation biologist and botanist Norman Myers, and at that rate, almost all of it could be gone by the year 2045. In Brazil, by government order since the late 1970s, rain forest "islands" have been left by ranchers clear-cutting for pastures, the island size varying from 2.5 acres (1 hectare) to 2,500 acres (1,000 hectares). In the 2.5–acre plots, for example, many species are soon gone: army ants and other insects that range over large areas, most birds and monkeys, even many of the trees. A quarter mile away, in a 250-acre plot, many of these species would survive. But certainly not jaguars. To save jaguars and to keep the number of disappearing species to less than 1 percent, areas of more than 1,000 square *miles* will be needed. For now, the Brazilian government has designated a few such areas as reserves, but in a fast-growing population and a developing economy, future demand for land threatens them. The "Minimum Critical Size" project—now part of the larger Biological Dynamics of Forest Fragments Project—will continue for the foreseeable future to monitor and recommend reserves of an optimal size and shape to protect as many of the species as possible in forest ecosystems all over Brazil.

MacArthur never dreamed that the work in Brazil or that practical experiments of his work would happen on such a grand scale. When Wilson talked with MacArthur in his last hours, they spoke keenly about the unsolved problems of evolution and the future of ecology; according to Wilson, MacArthur talked as if he would live forever. Biogeography, as part of ecology and population biology, has become important to evolutionary biology because it illuminates the structure and organization of animal and plant communities over time and space. Yet neither MacArthur nor Wilson realized then how vital their collaborative enterprise would become—not just for the ants and the tropical rain forest but for conservation in general.

Since his visits to the tropical rain forest in the Amazon and Costa Rica in the early 1980s, Wilson has become one of the leading proponents for the study and conservation of biological diversity. He began to write critical essays in the scientific and popular press alerting the scientific community as well as the public to the "biological diversity crisis"—the loss of biological diversity due primarily to deforestation in the tropical rain forests. Wilson, with several colleagues, attempted to measure the rate of loss of species. In the early 1980s the rate was thought to be about three species a day—one thousand a year. But by the early 1990s Wilson's revised estimates, based on a conservative figure of 10 million species living in tropical rain forests, indicated the extinction rate could be as high as seventy-four a day—twenty-seven thousand a year. Many of those species are unstudied, even unnamed.

Since Linnaeus first developed the binomial nomenclature system in 1753, some 1.4 million species have been catalogued and named: 344,000 plants, including algae and fungi; 42,000 vertebrates; and about 750,000 insects. The remainder of about 263,000 species are other invertebrates (arthropods) and microorganisms. More than half the known species are insects. But the real numbers of insects are much higher. As Wilson and others have begun to explore deeper in the rain forest, they keep finding more diversity. The latest frontier is the canopy, the highest level of the rain forest, where temperatures soar to 120 degrees Fahrenheit, and water is hoarded in the leaves and stems or brought from below, sucked up through the sort of aerial roots that Tarzan used to enjoy swinging on.

Until recently, the canopy was considered as inaccessible to humans as the surface of the moon. In 1982, in one of the first detailed canopy studies, Terry L. Erwin of the U.S. National Museum of Natural History, working in Peru, fired canisters of insecticide to try to find out what might be up there. Thousands of insects from a single tree fell to their deaths on sheets laid out on the ground. Erwin, a beetle specialist, started counting new species and then lost count. Extrapolating from the "canopy fogging" work, however, Erwin suggested that there might be as many as 30 million arthropod species yet to be discovered and named, mostly in the tropical rain forest canopy and mostly beetles. This estimate has come under much debate, with some biologists opting for much lower estimates of 3 million to 5 million undiscovered species and others, particularly entomologists such as Wilson, saying it's not high enough, that 10 million to 100 million species may be closer to the real figure for life's diversity. Whichever, the numbers are astonishing. The bald fact is that we still don't know even to the nearest order of magnitude how diverse life on our own planet might be.

To Wilson and Brown and others who make pilgrimages to the natural world, the tragedy of lost diversity really hits home. They return to a favorite area to see old ant friends, and it's a clear-cut farm, a one- or two-tree species second-growth forest. It has happened more than once to both men. The tamed landscape is almost silent, no whoops and whistles of birds, no monkeys, none of the teasing complex cacophony of cicadas and other insects. The leafcutter ants, however, are probably still there. In general, ants are among the last to go. Ants and many other insects will in time adapt, but the diversity of ants, especially those narrowly adapted species such as the dacetine tribe, will almost certainly decline. But before that, most of the larger fauna, the birds and mammals beloved of many humans, will disappear.

As part of his increasingly public profile in recent years, Wilson has tried to promote a return to field research and taxonomy—to find and evaluate species before they are lost forever. The loss is not just to science, as Wilson says, it's a loss to all of humanity. A loss of possible new medicines, foods, and other products and opportunities that we will never know. Nearly half of all medicines are based on natural products; wild relatives of crop species are crucial for protecting crop plants from pests and diseases and for improving tolerance to drought, salt, heat, cold, waterlogging, and many other conditions. "Biological diversity is quite simply our capital, our source of wealth," says Wilson. Nature is also our main source of wonder on Earth—and a spark to the creative imagination.

Wilson and Brown's love of biological diversity goes back to their jackdaw boyhoods, to all the natural treasures they hoarded. Both collected numerous things. Brown, growing up in Philadelphia, collected "everything, even streetcar transfers." His first living collection was plant specimens, mostly from trees. He was astonished at the diversity. Wilson's first great collection was snakes. Fascinated by them, his goal in his mid-teens was finding and catching one each of the forty or so species of snakes found in the southern Alabama–Florida panhandle area. He very nearly did it. At Boy Scout camp one summer he had his own mini-zoo and would lecture on each species. By fifteen his reputation was such that he was nicknamed "Snake" Wilson at school. A bite by a pygmy rattler with an emergency run to the hospital was not warning enough: He challenged a mature five-foot poisonous cottonmouth in a reckless solo attempt to capture the monster of the swamp. Fortunately, he let go in time, heaving the snake away from him.

Wilson was an only child. His father was a federal accountant through the Roosevelt era who moved around a lot, mostly to small towns in Alabama and Florida, with a short period in Washington, D.C. Growing up, Wilson managed to attend sixteen schools in eleven years. Wilson's parents split up when he was seven, and for financial reasons Wilson was raised largely by his father and stepmother, although his father was an alcoholic who later committed suicide. In any case, Wilson kept in touch with his religious, sensitive mother, visiting her and his stepfather every year while he followed his father's "long and complicated odyssey." As he changed schools and friends and circumstances, he found his solace, in the woods, never more than a bicycle ride away. Even when Wilson lived in Washington, D.C., Rock Creek Park did the trick. He didn't mind being a loner, but he usually managed to conscript someone to help in his expeditions to the woods. The role was always the same: part-time zoologist and chief nest excavator. Had Brown and Wilson met before Harvard, they would have gotten along famously. The only question is who would have done the most digging.

CONVINCED HE HAS the find of the century, or at least of the day, Brown is grumbling at full volume, heaving and straining to support a mighty log presumably to be carried into the work area. He is looking for his chief nest excavator. He calls Wilson for help.

"Bill, good God!" says Wilson, generally patient but having his patience tried. "You'll never get that log up on the table."

"We'll do it *in situ*," says Brown. "Give me a hand."

Wilson takes the other end of the log, and together they rock it once, twice, until it gives, rolling over with a thud to reveal, on the underside, a few hundred seething bodies of various ponerine ants, sabers drawn. The acrid stench of ant battle is in the air. As if to join the fray, Brown steps back and pulls out his own shiny, long blade from a neat leather holster. The machete slices through a bit of sunlight streaming through the trees and flashes briefly.

"Where did you get that?" asks Wilson.

"Borrowed it from the kitchen," he says beaming. Brown starts hacking out prospective ant hideouts, ignoring Wilson's gibes about its being "a girl's machete—the holster probably has a macramé belt." And Wilson, too, joins Brown on his fantasy log—this rotting, ant-ridden home to various scavengers, killers, and thieves—which to Brown's eye just might contain all manner of special things.

"The miracle ant, Ed," whispers Brown.

"What!"

"*Thaumatomyrmex!* Isn't this your miracle ant?" says Brown. "Wait a minute, hold the phone."

Brown is teasing him, but then, just for an instant, the light catches the shiny black back of a solitary hunter, and Brown thinks, well, maybe it is. But after a closer look with the loupe, both determine it is not.

"Well, close, Bill," says Wilson. "But no prize."

"Here's one of those tiny *Pheidole,*" says Brown.

Brown pronounces it *fay-doley,* Wilson *fye-doley. Pheidole* ants are the predominant genus of ants, the background fauna, throughout the world—especially in the tropics. The carpenter ants *Camponotus,* also worldwide, including on many islands, are often described as the most predominant with perhaps five hundred species in the genus, a mess for taxonomists. But Brown and Wilson know that when it comes to ant diversity, the genus *Pheidole,* outdoes them all. There may be one thousand to two thousand *Pheidole* species in the world and at least six hundred in the New World alone.

"Are you going to take it back?" asks Brown.

"We have so many tropical *Pheidole* we're drowning in them," says Wilson. "I have such a backlog of specimens to classify. My God, we have a zoo here, Bill. Oh, this place is swarming with ants! The miracle ant could show up anywhere here."

"Okay, I'll take it back," says Brown. "Are we ever going to finish that monograph?" They have agreed to try to come to grips with the huge *Pheidole* genus and write a monograph that will tell the tale of how this dominant genus became so well established and then radiated throughout the world. It could be the ant monograph to end all ant monographs; no one dares guess how many pages long.

"I don't know," says Wilson. "We'll probably go out on *Pheidole.*"

"After we finish it," says Brown optimistically. "We can do all the rest of the Myrmicines." The Myrmicines include all the leafcutters, the fire and little fire ants, as well as *Pheidole*—about half of all ant genera. They are the largest and most disordered of *all* the subfamilies of ants.

"Do you remember that movie," asks Wilson, "the one with Paul Newman as the ultimate cad and Patricia Neal as the worldly wise woman? Was it *Hud*? He's coming on to her, and she says, 'I've already done time with one son of a bitch, I'm not about to take on another!' Well, when I've finished *Pheidole,* Bill, I've done my time."

"Ah, right. You have got something running on your arm," says Brown.

"My God, this place is alive with ants!" says Wilson, not paying attention.

"See, it is playing dead now," says Brown.

"Another *Pheidole,* different species," says Wilson, finally noticing it. "It's tiny, and here's the colony it belongs to; it's just minute. I'll bet that a lot of these *Pheidole* are small because it's an escape mechanism from the army ants that prey on them. It's like pygmies. Army ants and other big ponerine ants can't get to them as easily; they're not worth the trouble. So dwarfism may have survival value."

"But when you get down to that level," adds Brown, "you're probably vulnerable to other enemies."

"We're already up to eight species," says Wilson, "and we haven't even started moving along the log."

"I'm going to dig some more out with the machete!" says Brown.

"Wait a minute," says Wilson. "Here's something walking along carrying a beetle larva. She went out, caught it, and now she's bringing it home. Uh-oh, it's a *Pachycondyla,* Bill, one of your favorite ponerines that stings like hell."

Bill Brown's "favorite" ponerine ant has characteristic yellow tips on its antennae, which it moves around like a wasp. No one knows why it does this; a few other ants do it as well. *Pachycondyla* are fierce, well-equipped ants that live in colonies but always hunt in solitary fashion.

"The trick is to get her by the abdomen and hold on to her before she grabs you," says Brown.

"She's all yours, Bill."

"Or you can hold her by the legs in such a way that the tail end and the sting are over your thumbnail," adds Brown. "Like this. Ouch! Well, it usually works."

"You'll make me a believer in big logs," says Wilson, ignoring Brown's cry of pain. "Here's a *Brachymyrmex.* See the small yellow body. It's one of the smallest ants in the world. Remember that *National Geographic* photograph of the smallest and largest ants in the world? *Brachymyrmex* was on the antenna of the big one."

As the tension over their argument at lunch dissolves, Wilson and Brown are truly back in their element, quietly digging away, exchanging occasional jabs and retorts, happy as two kids prospecting in the woods with nothing to do but get dirty, maybe get stung by something interesting, and see what the day might turn up. But wafting up from the earth is a smell, described as meaty by some and like potato blossoms to others.

"Do you smell anything?" asks Brown.

"No, Bill," says Wilson. But after a few seconds, he looks up. "Oh, I see we have visitors."

A long column of ants is approaching their prize log and about to climb over it, under it, all around it, and into it in search of ant colonies and other odd insects. It is a solid tangle of rugged marching ants. These are the Huns and Tatars of the insect world, as William M. Wheeler once called them: the army ants. Of the various species of army ants, several of which are found at La Selva, these are of modest size and warlike disposition, a species called *Eciton hamatum.* These hunt only in raiding columns, rather than the intimidating massive fan swarms of their much larger relatives, the swarm raiders *Eciton burchelli,* which are the biggest spectacle of the Neotropical world, the ones that take anything that cannot get out of the way, including injured birds and small reptiles and mammals. *Eciton hamatum,* the column raiders, eating mainly immature ants and social wasps, invade other ant colonies for the juicy larvae and pupae. But both the swarm raiders *(burchelli)* and the column raiders *(hamatum)* "just raise hell with other ants," says Wilson, "and *Pheidole*—because there are so many of them around—are often the victims." All along their path through the jungle, the column raiders can quickly diminish the ant diversity of an area, although the impact on the diversity of the whole forest is minimal. Their name, *hamatum,* means "furnished with hooks" in Latin; the mandibles of the soldiers are hook-shaped. They also have strong tarsal hooks at the end of their long sturdy legs, which grip the ground firmly as they run and when they need traction to pull down big prey.

"This is probably an outermost branch of the column," says Brown. "Oh, here they come. They're really getting thick in here. They smell something on the table."

"They're on the march," says Wilson. "They might drive us right out of here."

"Ouch! We must have stepped on a head," says Brown. He tells Wilson when he "first worked that one out" years ago at Barro Colorado Island, Panama. One day while eating his lunch in the field, he decapitated an army ant worker and then put the body down. There was no reaction. When he squeezed the head and put it down, however, all the ants attacked. The alarm response signal is in a gland in the head, and the soldiers, which have the big ice-tong mandibles, have a superabundance of it. These mandibles are used to defend the ants from their mostly vertebrate enemies—anteaters, coatimundi, and the like—as much as for offense during their hunting forays.

Brown relates that one of the Amazonian Indian tribes showed him how they used the ants' needle-sharp, ice-tong mandibles to make sutures to close up wounds. They showed him a boy who had suffered a deep gash on his arm. They picked up a tong-jawed major worker, a soldier, and held the ant's head along both sides of the wound to bite the boy. In practice, they usually had to squeeze the ant's body between thumb and forefinger, which caused the jaw muscles to contract, piercing the flesh and pulling the open sides of the wound together. Then they pinched off and decapitated the ants, leaving one or more ant heads along the arm. "A couple had already fallen out," says Brown. "That's one of the nice things about them—you don't have to take any stitches out." The ant "stitches" disintegrate and the heads simply fall off.

American behavioral psychologist turned myrmecologist, T. C. "Theodore" Schneirla, who made a dozen expeditions between 1932 and 1967, to study the army ants of Central America, found the practice prevalent and of long standing in certain interior regions throughout tropical Latin America. Apparently, several species of the larger army ants, including the column raiders and swarm raiders, and even a few other ant species, are used in this way.

"It's an authentic native remedy," says Brown.

"Here's another feeder column," says Wilson. "This colony must have a couple hundred thousand workers."

"It's a good thing we don't have swarm raiders right now while we're excavating this log," says Brown.

"Oh, we'll meet them, Bill. Just you wait."

As the army ants concentrate their forces, making a siege on the prize log held by Wilson and Brown, the two men bag a few more samples and stand up as if finishing an enjoyable if slightly large meal.

"Well, Bill, there are probably twenty-five ant species on that log," says Wilson.

"That's a real smorgasbord for the army ants," says Brown. "Shall we leave them to it?"

Walking down the path, Wilson and Brown refocus their eyes for miracle ants. It may take a miracle worker to find one.

"This is near where I once saw that primitive dacetine—*Acanthognathus,*" says Wilson, "so we may turn that up. That would be a thrill."

"I remember when I found that first one in Colombia," says Brown. "I thought it was carrying a big stick in its mouth. An ant carrying a stick! God, what in the world is going on here? The mandibles were so long, and they were open a full one hundred and eighty degrees."

As they collapse in late-day punchiness, Brown and Wilson have the sense of humor to realize that their specialty is at least as arcane and esoteric as the business of a hunter ant seizing a springtail or a beetle larva and carrying the treasure back to the colony. This is not to say that ants are arcane and esoteric. Ants are the stuff of this Earth, more than most groups of animals and certainly more than humans, even if it's up to humans, myrmecologists, to unravel it. Of course, myrmecologists, as scientists go, are close to the earth, if not "of" it. And they have the good sense to know when it's time for supper and a beer, time to call it, in Wilson's words, "an honest day's ant work."

Packing up their vials, bags, and boxes, portable chairs and table, they begin the trek back to the bunkhouse. Passing the log where they had found so much ant diversity, Wilson notes that the column-raiding army ants have started to retreat as well, carrying their booty, as they typically do, after foraging most of the day. As if to fill their place, a little farther down the path, thousands of miniature army ants, *Neivamyrmex,* are moving out of their underground nest, their antennae bent downward, just above the trail. Compared to the column raiders', the torso of the mini army ant is much smaller and more slender, and with shorter legs—adaptations for their more underground habits. Their underground nests are sealed shut by a wall of ants, each ant with her legs intertwined with the next. Along with the column raiders and the swarm raiders, these miniature army ants are members of the New World army ant tribe Ecitonini, but they belong to a separate genus, *Neivamyrmex,* the largest with some 117 species. This genus is typically found in the more difficult or marginal areas that army ants inhabit: in deserts, cooler temperate areas, or underground in various forests throughout the Americas from as far north as Virginia and Iowa to southern South America. In only a small portion of their range, mostly through Central America, do they overlap with the large, classic army ants. In Costa Rica, mini army ants typically start hunting late in the day and work the graveyard shift.

The light is fading and darkness hits the tropics abruptly—no time to linger just now. Still, no harm in looking for the miracle ant. For the half-mile walk back to base camp, both heads remain resolutely bowed.

Chapter Three

Superorganism

The observer who first comes upon these insects in some tropical thicket [senses] a subtle, relentless, and uncanny agency, directing and permeating all their activities. . . . Like [a] person, it behaves as a unitary whole, maintaining its identity in space, resisting dissolution.

—*William Morton Wheeler,*

ANTS, *1910*

A solitary ant, afield, cannot be considered to have much of anything on his mind. Four ants together, or ten, encircling a dead moth on a path, begin to look more like an idea. [But] it is only when you watch the dense mass of thousands of ants . . . blackening the ground that you begin to see the whole beast, and now you observe it thinking, planning, calculating. It is an intelligence, a kind of live computer, with crawling bits for its wits.

—*Lewis Thomas,*

THE LIVES OF A CELL, *1974*

There are . . . many ways in which ants and human beings are alike. Both are resounding success stories of evolution . . . and both have achieved their success through their ability to form social groups, to communicate, and to manipulate their environment with great dexterity.

—*Edward O. Wilson,*

FROM AN INTERVIEW, *1990*

The Advancing Arm of the Superorganism. Army ants prowl through the jungle, their "ice tong" mandibles ready to strike.

By night the closed canopy of the tropical rain forest may be the blackest place on Earth. Even in the moonlight, the lush foliage, so dark and fleshy, allows scarcely any light to pass through or be reflected. Yet no matter the depth of the darkness, the rain forest hums with assorted low chirps, grunts, whistles, screeches, thumps, hisses, and sighs—so much rustling in the leaf litter, scrambling on trees, winged beating of the dank, heavy air. But even more than sounds, there are smells, as creatures go about their business of moving and communicating, hunting and fighting, loving and dying. Every behavior in each of many thousands of insect species produces a different chemical scent composed of a stream of airborne molecules rich with meaning. With little or no wind, the molecules hang in the air for minutes, waiting to be picked up by insect antennae and translated into action. Other scents may cling to the earth or trees for a few days. Almost all of this nighttime insect world resides in this chemical realm. Even so, many insects of the night—including ants—tend to be big-eyed, eager to exploit any advantage they have. There is evidence that certain insects can orient visually at night and that they sometimes even navigate by moonlight. Vision is also important for flying insects—especially when mating or trying to find a mate. But insects would be solitary, lost creatures without chemical communication. It suits a small-brained creature that depends on instinct. That's why the insect way is Better Living Through Chemistry. This is a world closed to birds, most of which have no sense of smell. Mammals do employ olfactory communication, but it is a different world from that inhabited by insects. The insects' chemical world is closed to the primates, various spider and howler monkeys, who sit out the night curled up with family high in a rain forest tree, or to those other primates who sleep a few hundred yards away in the station bunkhouse, waiting for daylight to return.

Among those creatures that live by night, the miniature army ant colony, one of the *Neivamyrmex* species, crawls out of its underground hideout at the first sign of darkness. Its makeshift home, as with other army ants, is composed mainly of the ants themselves, which interlock their legs together to form the walls. Army ant workers are nearly blind, with two degenerate compound eyes that are sensitive to changes in light but not much more. But the miniature army ant colony responds quickly to light's absence, knowing it now has its turn to forage through the jungle. In this way it avoids meeting the larger army ants, the column and swarm raiders that operate by day. Occasionally, however, the miniature army ants will forage on an overcast day, fooled by the lack of light. And they come out earlier in the darkest parts of the closed canopy rain forest. But in these places, the column and swarm raiders, also regulated by light, often retreat early, returning to their nests and leaving the miniature army ant colony to its dark prowling. For this colony, night is a niche, its best bet of carrying on in the face of competition with other army ants.

The nighttime world of ants has been recognized, if largely left unstudied, for many centuries, even going back to Aristotle. In *Historia Animalium,* Aristotle wrote that their architectural labors occurred not only during the day but when the moon was full. The Swiss myrmecologist Pierre Huber, writing in the early 1800s, affirmed Aristotle's findings but realized that ants work on moonless nights as well. Huber, too, by identifying individual ant species, began to see that ants had different working rhythms depending on species, location, weather, and other factors.

As the colony of some eighty thousand mini army ants starts foraging, it branches into several long lines and streams out into the jungle night. The colony marches in columns, as do most army ants except for the big, lumbering, long-legged swarm raiders that fan out in wide swarms across the jungle. It is easier for insects to avoid a column of ants than a massive swarm raid. Still, to small ant colonies, termites, and other insects, these mini column raiders can be relentless. Moving slowly but steadily, they will grapple with anything that cannot flee fast enough—even animals several times their size. It does not matter that they lose many of their own in the process—sometimes to beetles, spiders, or other larger ants. Their secret is in the sheer size of their army. Working together, several will first seize the prey in their mandibles, subduing them, while their abdomens curl around the victims to inject the poison. The mandibles are well equipped for macerating the booty. The nighttime world hears, feels, *smells* the threat. The question is whether to stand one's ground —to fight for one's colony or food resources and risk being overwhelmed— or to flee.

One ant that is cautious about the advancing army ant colony is the leafcutter scout. The scout has shifted her foraging from day to night, as a leafcutter colony can do from time to time. The shift is often due to competition from some other ant colony. When two leafcutter colonies forage in the same area, alternate shifts become practical. Or if the daytime raids of column- and swarm-raiding army ants start encroaching too much, then foraging on certain trees or bushes may be best left to the night. But inevitably the leafcutter ants are so numerous and widespread that some run afoul of army ant raids and any number of other hazards such as birds, toads, lizards, and anteaters. This is life in the lowland Neotropical jungle. But leafcutters, with their rugged, spiny torsos, are not especially appetizing, and their well-equipped soldier force discourages most predation. For the most part, researchers have observed that army ants and leafcutters manage to avoid each other.

Besides the leafcutter scout, there are four bullet ants remaining from the massacre of their colony. Without their queen or colony, they are living on borrowed time. They have been returning to the empty nest by day and foraging high and free in the canopy by night. They are workers programmed to work, but without their queen, there is no plan and no future, just these thin, individual lives stripped of meaning. Yet for now they carry on.

The leafcutter scout is the first to respond to the danger of the advancing miniature army ants. She runs back toward her colony at the edge of the forest near the overgrown plantation along the Río Puerto Viejo, avoiding the columns of the miniature army ant colony. Just thirty seconds later the bullet ants have lingered too long. As the army picks up the heavy, musky, unmistakable odor of the bullet ants on the antennae of its workers and soldiers, the message spreads quickly through the column: Fresh meat! Standing four abreast, the bullet ants open their mandibles and hold their massive black heads up high, their thick gasters flexed to make the stings point forward threateningly. It is an awesome sight. Four elephants preparing to charge. The bullet ants begin to squeak and click loudly, aggressively. The four elephants are soon surrounded by thousands of pygmies.

The bullet ants squirt out heavy doses of musky pheromones—their alarm signal. The molecules suffuse the nearly still night air. But there are no other colony members to receive the message. The four balas are alone. With no colony to process their signals, there will be no sister workers running to assist. Meantime, a few hundred feet away, the leafcutter scout has rejoined her colony. Squeezing out her own alarm pheromone, the message is quickly passed to the main trunk line and from there to the colony. The big leafcutter soldiers are dispatched instantly to guard the plant-foraging arms of the leafcutter colony. They will be ready to assist with a strategic attack or a large-scale retreat if necessary.

But the bullet ants are not so fortunate. As about twelve miniature army ants grab the legs of the first bala and squeeze shut their mandibles, the big animal lurches in pain, snapping its sharp jaws and killing three miniature army ants in one mandibular stroke. Molecules of army ant alarm pheromone are sprayed all around, suffusing the air. The other bullet ants start squeaking loudly as they prepare to sting some of their attackers. It turns out to be a trumpeting before the final hour.

The miniature army ants crowd around, pushing in tight, all seemingly eager for a piece of the action. Nine of them surround and take turns snapping at the first bala. She arches her abdomen, sting at the ready. They spread-eagle her and begin gnawing away. Then they sting the bala on the abdomen, subduing her with their poison. The big female flips on her back, flinging half a dozen little workers in the air. Simultaneously, the three other bullet ants begin making their moves. One charges into the fray, mandibles flailing, her antennae folded back into the scrobes, or grooves, in the side of the head to protect them, her stinger curled up in the air, at the ready. It is a kamikaze attempt to inflict the most possible damage—as if she still had a colony to defend. As she slashes three of the tiny ants, stabbing abdomens, decapitating one and mangling the legs of the others, the mingled smells of battle, the chemical cries for help, ruin the air. The two remaining bullet ants turn and run, slipping away on a half-fallen branch still connected to the trunk of a mature balsa tree. Hanging by a shard, the branch forms a bridge to the trunk of this massive tree that extends all the way to the canopy. The balas scamper across the bridge and onto the trunk and head up toward the canopy. One bala suffers a twisted antennae where an army ant momentarily grabbed hold of her, but both ants are otherwise unharmed.

Down on the ground, the alarm pheromones now summon the larger miniature army ant workers who serve as soldiers when needed. The smaller workers step aside while the slightly bigger, clumsier soldier-workers muscle in to finish off the two balas. The death throes ensue, and it is over in a few seconds.

The soldiers pursue the two escaping bullet sisters but soon give up. Fleeing, the sister ants continue climbing the tree, the tarsal claws on their legs gripping the smooth gray bark. The army ant colony, which often climbs trees, is not as fast as the long-legged bullet ants. In a stroke of luck, the lead bala picks up her own trail pheromone on the tree that she had laid hours earlier to some active, rich extrafloral nectaries where she had found the liquid sugar that many ants love to drink. In balas, the trail markers are unique to each individual ant rather than common to the colony as in most foraging ant species. Balas don't normally recruit other balas to food but bring food back to the colony. The bala with the twisted antenna follows close behind, orienting by touching the gaster of her sister with her one good antenna. Following the previous trail, the last surviving bala ants press ever higher, skirting lianas and strangler figs, a termite nest, and a colony of ants that mimics termites. She avoids the bright orangish red and dark blue poison dart frogs, *Dendrobates pumilio,* probably wary of their skin secretions. They are related to the frogs used by the Chocó natives of Colombia to supply poison for their blowgun darts. The alkaloid poisons secreted from glands in the skin make all poison dart frogs highly toxic to animals, but it is not known how toxic they are to insects.

A few careful steps later, the two ants hear a piercing "tink." It comes from the tink frog, one of the *Eleutherodactylus* species, hiding in a bromeliad plant on a branch. The loud sound is out of all proportion to the frog's size—an inch long, about the same size as the poison dart frog and barely larger than the bullet ants. The tink frog is cryptically colored and lacks the toxic skin; hence its retiring nature. In experimental research at La Selva, bullet ants were presented alternately with poison dart and tink frogs. The balas refused the poison dart frogs, either immediately or after biting them and repeatedly wiping their mouth parts on their front legs and on the tree bark. But the tink frogs were almost always taken, stung, and killed. Both of these frogs eat ants and other insects—although balas are a bit big. Yet today there is no contest. Neither the bala sisters nor the frogs are interested in tangling with each other.

At the first large branch in the tree, at thirty-five feet, the balas find some rotting figs and, inside one of them, fresh moth larvae. These may have been left from the bats that pollinate balsa trees, or perhaps this branch was used as a temporary waystation by one of the toucans guarding the nearby fig trees. The balas share pieces of one fig, biting into the moth larvae and pulling them apart with gusto. They then continue climbing higher until they are far away from the smell of the events on the ground where the miniature army ant colony is negotiating the transport of the fresh meat from the two dead balas. Ten or twelve army ants surround each bala carcass and, with their mandibles, grab hold of a leg, struggling to lift the bodies off the ground. Finally, the army ant workers start dragging them back along the foraging column, back to the main part of the colony. There is fresh meat now for the larvae and the queen, fresh meat for those who need it. It does not matter that a few dozen army ants have died in this mini battle. The colony overall has gained immensely from the food kill. Then the column goes back to its probing, to its wriggling advance through the nighttime jungle. There are a thousand smells and tastes all around to be investigated. There will be food for all. From high above, the last bullet ants—the last survivors of the bala colony—pause from time to time on their trek to the canopy. They cannot see the advancing armies below. Only occasionally, as they continue to climb, do they catch a whiff from the scene of destruction below.

The battle between the bullet ants and the miniature army ants is a cool demonstration of a colony's power. If the bullet ants had formed a colony, even though it would not be as numerous as the army ants, it could have held off the army ant colony, perhaps forced them to back off or change the column's approach. But individual bala ants, no matter their size, have no defense against a large colony of attacking ants. The battle is not between individuals but between a single organism of massive size, resources, and capabilities and a colony fragment made up of several individual ants: That is no contest at all. Had the miniature army ants attacked the leafcutter scout, for instance, things would have been different. The leafcutter scout might lose her life, but she would die in a massive cloud of alarm chemicals that would serve as urgent marching orders to the soldiers of her own colony, as large or larger than most army ant battalions. The leafcutter colony can hold its own with almost anything in the jungle, although sometimes, such as when army ants are on the move, the strategy is simply to wait underground until the fighting has passed. Of course, if the scout's pheromones are too far from a main trunk line, then the message never gets relayed, or it may be picked up days later when it's too late. When she or other scouts are killed too far away from the foraging trail, as sometimes happens, then the colony is effectively blinded. There are ant colonies that regularly knock off scouts from nearby colonies just to keep competing colonies blind—either to protect food sources or the colony itself. As in any conflict, preemptive action can be the best strategy.

The column of army ants is the probing arm of a hungry organism. It mainly grabs the adults and brood of various small ants and termites. But when it encounters a large individual ant such as a bala, which offers substantial meat, the colony harnesses its forces: Larger workers, or soldiers, are dispatched, and the grasping claws of the colony turn from functional gathering tools to sharp weapons as the prey is subdued and killed. Then the probing "arm" transports the meat to booty caches and from there to the main part of the colony, the bivouac, which is the portable army ant nest. The bivouac is often situated beside or beneath a log, and its walls are made of living ants. Their outstretched intertwined legs create the walls of a shelter for the queen and her massive brood. There, the meat is fed to the queen and the colony's brood and shared among the workers. The miniature army ant column is not a single ant, not a hundred, not a thousand, not ten thousand, but, to paraphrase Lewis Thomas, it is the arm of a single writhing, ruminating beast. It is a *superorganism.*

Now the miniature army ant superorganism approaches a little fire ant colony, *Wasmannia*—a fairly young, small colony. The little fire ant, too, scavenges but can also fight. The ant with the longest reach, or foraging columns, and the sharpest claws, or largest soldiers, is often the victor. But sometimes it's just a case of how fierce or how hungry the ant is at the moment of attack. First the mini army ants penetrate the little fire ant middens, carrying off the pieces of odd ants, beetles, and other food scraps. Meanwhile, four little fire ant workers each pick up a larva in their mandibles and run for the tunnels in the wood. The miniature army ants scamper after them and are just small enough to squeeze below. Seconds later when they reemerge from the tunnels, the miniature army ants, obvious victors, have captured the fresh, still wriggling larvae, while other miniature army ants sting the workers and carry them out dead.

The little fire ant colony has had an arm severed, but it still lives. For the moment, the queen and a small retinue of workers remain hidden in a deep cavity in the log. The queen with hundreds of eggs inside her has the means for continuing her species if she can only keep a few workers to bring her food and to help with rearing the young. Yet no queen is the leader or the brains of the colony. The brain of an individual ant is composed of at most a million nerve cells, compared with 100 billion or more neurons forming the human brain. The superorganism's brain is the entire society, and in a mature colony of the leafcutters, that could mean 3 trillion neurons—thirty times more than a human. With each ant performing a number of different behavioral acts, the caste system and division of labor create a complex yet effective colony unit. Even if the queen is not in charge, she is the most important part of the superorganism, the single individual that no colony can be without. The queen is the colony reproductive organs. In some ant species, such as leafcutters, the queen is able to live for two decades or more. But when she dies, no matter the size or species of ant superorganism, the whole beast dies. Having more than one queen, as some ant species do, is added insurance, of course. It's more difficult to eradicate multiple queen colonies. In some parts of the world, little fire ants live in multi-queen colonies that dominate the insect fauna of entire islands.

The mini army ants circle the little fire ant colony, poring over the remains, fighting drunk on the odors. The workers search the passageways until they find the queen. Some return almost immediately to recruit more workers, laying their own pheromone trail. One by one the army ant workers destroy the last loyal workers attending the queen, and then they surround her. Even though the queen is larger than the workers, she's still extremely small—less than a quarter-inch long—smaller even than her assassins, the "pygmy" mini army ants. Her body, ill-suited for defense or food collection, is specialized in reproduction. This young fire ant colony has but one queen. In other parts of their shared range, where fire ants live in multi-queen colonies, little fire ants might prevail over the mini army ants. But not here, not today. As the fatal stings are administered, the colony is effectively rendered infertile, just as the bala colony had been a few days earlier.

A few minutes later, another "arm" of miniature army ants, a new column, has started to scale a fifty-foot-high cecropia, or trumpet, tree (*Cecropia* species). This common Neotropical representative of the mulberry family grows rapidly, colonizing light gaps, forest edges, and stream banks. Two distinguishing features are the bamboolike rings on a gray trunk and the large, deeply lobed leaves, whitish underneath, that resemble parasols. Cecropias are some of the most popular trees across both dry and wet tropical forests. One researcher counted at least forty-eight animal species, including leafcutter ants, iguanas, mammals, and especially birds, using cecropias for food, housing, or other purposes. In a single piece of Central American forest, American Neotropical biologist John C. Kricher found thirty-three species of birds visiting cecropias. The assembled birds make a colorful show when the trees produce their fingerlike catkins. When the catkins start to turn to fruits, the birds squabble over them with various mammals from bats to monkeys. To these and other tropical animals, cecropia trees are fast-food restaurants.

But not all cecropias are "open for business." In some, customers receive a painful welcome. This particular, nearly mature cecropia has its own standing army, legions of the fierce aztec ant *(Azteca).* It is an ant colony in residence. Not only are the approaching miniature army ants refused access to the galleries in the hollows of the tree trunk—strictly reserved for aztecs—but the aztec soldiers will not permit any other insects even to crawl on the tree. If they start to climb on the trunk, they do not get far. Birds and quick mammals such as bats still nip in and take fruit from the tree, but they cannot linger. These highly aggressive ants even patrol the trunk and extensive branches, sweeping off the seedlings of epiphytic plants—plants that grow on other plants and mooch water, sunlight, and sometimes nutrients from an existing tree. And they sometimes patrol the area around the trunk, attacking ants or other animals, including humans, before they approach too close or even think about touching the tree. In "repayment" for these services, the tree houses the colony in the hollow trunk and branches, and provides some food. This symbiotic association makes a strong fortress.

The miniature army ant colony is soon repulsed. Shortly after the miniature army ants start to snake up the tree, they have to pull back, as if slapped on the knuckles, and a few dozen individual ants are killed. The aztec ants have no stings; the mini army ants are bitten and then sprayed with noxious chemicals —or simply pushed off the branches. A few aztecs die in the skirmish, too, but the dead on both sides are just sloughed skin, a few dead cells. A worker ant of an army ant colony may live only a few months anyway before she is replaced by another worker, and another and another. The worker parts are replaceable, interchangeable. And now the colony reaches out in a new direction, toward the log where the myrmecologists had found so many different ant organisms the day before.

Harvard's William M. Wheeler first compared ant colonies to organisms in a 1911 paper in the *Journal of Morphology* which was widely read by biologists as well as philosophers. In "The Ant Colony as an Organism," Wheeler wrote that colonies of ants and other social insects are "true organisms and not merely conceptual constructions or analogies." He pointed out that a colony maintains its identity when foraging for food and defending against intruders. Each colony has its own life cycle, its own parasites and diseases, and in some cases its own ant slaves or insect cattle that it keeps for work or for food. Each colony, moreover, is like a single biological entity, except when it produces a sexual brood, called the royal brood. Then, for a brief time, usually a month or two once a year, it becomes a tight matriarchal family, with a mother and her sisters raising virgin queens and male suitors before sending them out to mate and start their own colonies. Once they have left, the colony becomes a single biological entity again.

In their day-to-day lives, the sterile individual workers of a colony not only share the resources and protect one another but they are selfless about it. They are altruistic to an extent humans only rarely achieve, with each colony member willing to die for the overall good. All for one and one for all. And to a large extent, certainly within each caste but also largely throughout the colony, the ants are interchangeable. Like cells, if one dies, it can be replaced. And if one or more castes in an organism are removed, others will take up the burden until that caste is regenerated.

But Wheeler was troubled by certain things he could not resolve: "If it be granted that the ant colony and those of the other social insects are organisms, we are still confronted with the formidable question as to what regulates the anticipatory cooperation . . . and determines its unitary and individualized course."

A few years earlier, Count Maurice Maeterlinck—the Belgian poet, dramatist, and essayist who won the 1911 Nobel Prize for literature and ruminated throughout his life about insect societies—had asked essentially the same thing: "What is it that governs here, that issues orders, foresees the future . . . ?" In the first of Maeterlinck's three books on the lives of, respectively, bees, termites, and ants, he suggested that the "spirit of the hive" was the regulating force in social insects. Wheeler found this a poetic and appealing notion but not very scientific. Yet Wheeler himself, for a time, could come up with no theory or explanation for the factors controlling "the correlation and cooperation of parts."

Searching for the answer, in 1918, Wheeler adapted a concept from the French entomologist Emile Roubaud. Wheeler knew that every ant in a colony has a social stomach known as the crop, the first segment of the gastral gut, which stores liquid food used to feed other colony members through regurgitation. Wheeler also realized that the brood care by the adults was not a one-way affair. The adults bring food to the larvae, but as a kind of stimulus or reward, the larvae produce substances imbibed by the adults. Wheeler dubbed this exchange of substances between ants "trophallaxis," from the Greek words meaning exchange of nourishment. Later Wheeler enlarged the idea of trophallaxis to include the exchange of all chemical and other messages between the members of a colony. It was also extended to include exchanges between a social insect colony and other insects that sometimes live as "guests" in an ant colony. All had to communicate with one another, he theorized, and this was how they were doing it. Wheeler even had a theory, which has never been tested, that trophallaxis was the way that social behavior in social ants and other insects started. Unfortunately, Wheeler enlarged the trophallaxis concept so much that it lost all meaning. Modern researchers still use the word, but it is now considered simply the exchange of oral or anal liquid—the giving or taking of liquid nutrition. But there is no doubt that some chemical signaling occurs during trophallaxis, that it is not just an exchange of food but an exchange of information. In recent experiments with imported fire ants, researchers have tracked the rapid transfer of food and information through a colony, revealing how older larvae manage to communicate to the queen through the nurse workers who take the larvae's regurgitated food—in effect processed protein—to the queen who uses it in egg production. In what has been described as a "positive feedback loop," the larvae can thus communicate to the queen the optimum rate of egg laying for the colony.

In 1928, Wheeler started calling ant colonies "superorganisms," updating his original idea of the colony as organism into a less metaphorical, more productive concept. The superorganism, in Wheeler's words, was "a living whole bent on preserving its moving equilibrium and integrity." This was the first hint of the idea of colony homeostasis, defined by E. O. Wilson as "the maintenance of a steady state, especially physiological or social, by means of self-regulation through internal feedback responses." Wheeler's idea of the superorganism soon caught on and became hotly discussed in philosophical, scientific, and popular circles. Scientists such as Alfred E. Emerson, who wrote philosophical papers on social insects beginning in the late 1930s, adopted the superorganism as the perfect analogy on which to string factual knowledge of social insects and other phenomena. Emerson's insight was to see the superorganism as a unit of natural selection that had developed adaptations parallel to those seen in individual organisms.

But Wheeler, Emerson, and the entomologists of the first half of the twentieth century had barely an inkling of how extensive and important the chemical messages are to this superorganism. When Wheeler died in 1937, modern biochemistry was still to come and the word "pheromone" had not yet been coined. Pheromones are now known to play the central role in the organization of insect societies. Ants use various chemical compounds and an acute ability to taste or smell them in order to find their food, to navigate, and to communicate with each other. Through the late 1950s, most myrmecologists were aware of the chemical scents of ants yet still thought of them as simply nest odors and trail secretions used as signposts.

In the late 1950s at Harvard, E. O. Wilson launched his pioneer research into ant pheromones. At the time he kept in his laboratory a colony of imported fire ants, *Solenopsis invicta. Solenopsis* is the large worldwide genus of fire ants and *invicta*, this species, translates as "the unconquered one," an appropriate name as they have today spread throughout much of the U.S. South and as far west as Texas, displacing other ants and insects wherever they go. To Wilson they were a boyhood favorite from backwoods Alabama and the subject of his first publishable scientific finding; he was the first to note that they had arrived in the United States. A fire ant is only one-eighth of an inch long, but Wilson managed to dissect the rectal sac and the two main glands of its poison apparatus. The organs—barely visible to the naked eye—were washed individually, then crushed. Taking an applicator stick, Wilson spread tiny trails of each substance across several glass plates. With any luck, he thought, some of the ants might follow the artificial trail after he stimulated their appetites with a piece of food. When one of the glass plates was placed outside the nest, but before he could get any food ready, dozens of ants came streaming out. They ran the length of the trail but then milled around at the end, as if lost.

In one of his eureka moments, Wilson realized that the chemical could get the ants to go anywhere, food or no food. "Stretched out in a line," he later wrote, "the pheromone is not just the guidepost but the entire message."

The worker squirts the trail pheromone from what is known as its Dufour's gland through the sting in the gaster, laying it in a fine chemical trail. A single glandular reservoir inside an ant holds enough of this chemical compound to give numerous workers their marching orders. If one milligram—one-thousandth of a gram (or .000035 ounces)—were dispensed with maximum

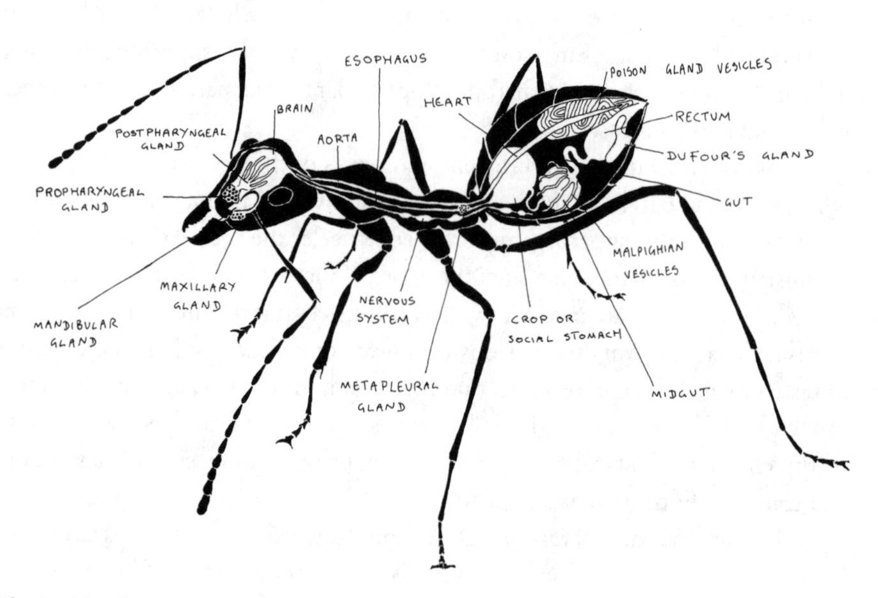

The Inside of an Ant. The ant is a walking battery of exocrine glands.

efficiency, Wilson calculated, it would be enough to excite billions of workers into immediate activity or to lead a column of them around the world three times.

Following his discovery, Wilson had a sleepless night. As he tossed and turned, he replayed in his head the Konrad Lorenz lecture of a few years earlier in which the Austrian ethologist had explained his theory of the visual and auditory sign stimuli, or "releasers," of birds and fish. One of the central concepts of ethology was that animals such as birds and fish are guided to a large extent by auditory and visual "releasers." The red belly of a male three-spined stickleback fish, a favorite study animal of early students of animal behavior, always produced the same complex behavioral response or behavior: another male stickleback would attack. So strong were the releasers that even a red spot on the underside of a piece of wood could stimulate the behavior.

Wilson saw that releasers could order and guide behavior for ants and other social insects, too, but in this case the releasers were chemical. In his excitement, Wilson dreamed of accounting for the ant's whole social repertoire with a few chemical releasers, each coming from a different gland and produced when an ant had to send a particular message.

Some months later Wilson found the alarm substances of fire ants in the mandibular glands. That's why when you crush the head of many ant species —by accidentally stepping on it, for example—you declare war (and sometimes a few other things as well): Soldiers will race toward you, mandibles snapping. As they mount their charge, the soldiers spread trail substance to call up more reinforcements.

Wilson also discovered the pheromone signal that tells one ant when another is dead. A three-day-old corpse accumulates necrophoric substances—unsaturated fatty acids and related esters. The odor tells the workers to remove the individual to the colony's refuse pile. But when Wilson painted ant-sized wooden dummies or pieces of paper with the substance, the ants carried them away. And when he dropped a bit of the chemical on a *live* nestmate, it, too, was instantly hauled off to the morgue, alive but unprotesting. After cleaning itself, it returned to the nest. But if not clean enough, it would be returned to the morgue a second or third time. Only when the cleaning was thorough would it be allowed to stay in the colony.

Since 1960, other researchers have found even more pheromones. The use of the word "pheromone" dates from the late 1950s, just after Wilson's pioneering discoveries in this field. Following Wilson's work, communication in ants has been studied intensively, and the results have produced a new understanding of social organization. The imported fire ant has been the most thoroughly researched, but studies of many other ant species have contributed to an astonishing, complex picture. Today, ants are known to communicate through the use of ten to twenty signals, depending on the species. A few signals are tactile or visual, but about 90 percent of them are chemical. The chemicals differ from species to species, and one species does not usually understand another's pheromones. Therefore, there may be as many different communication systems as there are species of ants—fifteen thousand or more. Even though the basic meanings, or use of the pheromones, are the same in many cases, certain ant species have evolved chemicals to convey unique messages.

Most of the pheromones are common to the workers and all other members of a colony. A few are specifically for communication from worker to brood, brood to worker, worker to queen, queen to worker, or from one queen to another queen. All these chemical signals are referred to as *pheromones* as long as the communication is between members of the same species. Within a single colony, if the colony can be said to be a superorganism, then the pheromones function almost as hormones. When the chemical signaling happens between two different species—interspecies communication—the name for the chemical substance depends on whether the communication is favorable to the emitter, an *allomone,* or to the receiver, a *kairomone.* Ants use such signals to enslave other ants or insects, or to pay deference to colonies or other insects that might capture them.

Most of the chemical signals come from glandular secretions. The typical ant worker is a walking battery of exocrine glands, as Wilson puts it. The exocrine glands produce chemical substances that exit through the mouth, anus, or other tiny holes in an ant's body. Three or four of the glands are in the head and the same number are in the gaster. In addition, there are one or two in the midsection. Most of the various glands are found in all ants, but their exact positioning and function varies from species to species.

When an ant leaks or squirts a substance from its head, it can mean a number of things. If it comes from the mandibular glands, it usually means alarm behavior, although some ant species issue alarms from the glands in the rear of the abdomen or gaster—either the poison, pygidial, or Dufour's glands. If the chemical comes from the gaster, then it commonly has to do with trail running. Some species use combinations of chemicals, while some combine a certain pheromone and a tactile signal, such as stroking of the lower lip or mouth parts, to get another worker ant to regurgitate some food from the social stomach or to get the brood to feed. Other times the message depends on the context. If the imported fire ant, for example, leaks Dufour's gland compounds during an attack, the response is attraction to the disturbed worker. Otherwise, the emissions from the Dufour's gland are for recruiting workers, and the response is for other ants to come to the signal and then to follow the trail of molecules. This signaling is employed in mass foraging and emigration from a colony. The strength of the message—the degree of danger, the number of workers needed for foraging, and so forth—can also be announced by the amount of compound used; the more compound dispensed, the more serious the situation is, and the greater the need for a fast, colony-wide response.

There are many other chemical signals that act as stimuli to ants. Special signature smells or territorial pheromones identify colony members. In some species there are also unique odors for individual castes and life stages, including eggs, larvae, pupae, and callow workers, or newly emerging adults. This helps workers identify and assist all the ants in a colony according to the particular needs of different castes and life stages. The queen often has her own smell. Special queen-recognition pheromones promote attraction to the queen and keep a colony together.

Every ant is a mini pharmacy of up to about ten compounds, with a different chemical in each gland. Examining numerous ant species, researchers have identified chemicals that include a wide range of compounds. A gland in one ant may carry an entirely different compound from the same gland in another ant. Closely related species, though, sometimes have the same or similar chemicals in the same glands. There would seem some potential for confusion, but closer examination of the pheromones reveals subtle chemical blending by many ants. Other components seem to be introduced to create signature blends unique to each colony.

In some cases, however, the compound used by one ant can mean one thing to one species and something entirely different to another ant or social insect. The lemony compound called citronellol serves as both the defense signal of the leafcutter ant and the aggregation—"let's get together"—pheromone in bumble bees. Four compounds that are sex attractants in various species of carpenter ants sound the alarm in Bill Brown's favorite *Gnamptogenys* and other related ants, and are used for defense in termites. It lends new meaning to the concept of "crossed signals" in the affairs of social organisms if "come and get it" can be mistaken as a declaration of war. But most of the time pheromones function extremely well. A particular pheromone is straightforward and exact in meaning within each species, and the response is hard-wired or instinctual in each ant brain.

The most complex system of communication known in ants, according to research by Bert Hölldobler and E. O. Wilson, is found in the African weaver ant, *Oecophylla longinoda.* Through a combination of chemical and tactile signals and unique ritualized patterns of movement, workers are able to convey at least five variations on the typical ant's call for recruitment: They can recruit workers to an immobile food source, especially sugary materials; they can recruit workers to new terrain; they can signal the colony to emigrate to a new nest site; they can signal short-range recruitment to enemies to provide for quicker capture of invaders and prey; and they can signal long-range recruitment to the vicinity of invaders and prey, used intensively during territorial wars with other colonies of the same species. These advanced weaver ants, which weave silk nests in trees, also have an elaborate pheromone-mediated alarm system.

How did ants come to manufacture, blend, and use all these compounds? And, an even more difficult question, how did ants develop the sensory organs and brains that enable them to read a cloud of molecules and respond to them instantly? Wilson and others have tried to answer some of the questions about pheromone evolution. The origin of pheromones can be glimpsed in behavior patterns present in solitary and "presocial" insects. Alarm substances are in many cases simply modified defensive secretions, such as plants use simply to deter predation, says Wilson. "And trail substances have a parallel in the odor spots that mark the nuptial flight paths of the males of certain solitary wasps and other insects." Trail communication in at least some ant groups appears to have evolved from "tandem running," a method of recruitment still practiced by some ants in which workers that have found food, for example, return to the nest and lead other workers out. Analyzing tandem running in a primitive ant species, Wilson found that the "leader" would advance two or three body lengths, secreting a chemical that excited the ant behind to drive forward and touch the "leader," which then pushed the "leader" to repeat the sequence. In all but the primitive ant species, pheromones have evolved to replace tandem running, and natural selection has shaped this behavior into trail communication—laying a trail for fellow workers to follow now or a little later on, with or without another ant to act as a guide. Yet tandem running remains as a tantalizing link to the past.

THE TWO BALA sisters, fifty feet above the ground, climb the balsa tree, heading toward the canopy. The hours-old pheromone trail that led to the promised extrafloral nectaries has turned faint, covered in part by a recent network of intricate trails laid by some busy weaver ants. But the weaver ants are nowhere to be found. The second bala sister occasionally loses her way, the twisted antenna presenting some difficulties. In the blackness she has to feel her way forward to her sister's gaster, touching it reassuringly with her good antenna. And because of the faintness of the pheromone trail, the first bala has to zigzag even more than usual, sweeping her paired antennae back and forth as she moves ahead, before losing the faint gaseous cloud of molecules on one antenna. Then she veers back in the opposite direction until she is back on course, before losing the scent again, this time on her other antenna. She must orient back until she encounters the faint cloud of molecules again. Yet once more for a few steps she remains in the cloud until she again slips out on the opposite side and needs to correct her bearings. It is a slow process, yet they make progress up the tree.

At the sixty-five-foot mark, they approach the periphery of a weaver ant's silk nest, constructed between various leaves and twigs of the tree. It seems to be in perfect shape but apparently abandoned. Dead sticks and dried leaves are pasted to the outer sheets of silk. A layer of thick detritus on the roof protects the nest from rain and direct sun, which sometimes penetrates the canopy. Highly evolved, weaver ants are relatively large and live in trees. They weave communal nests out of dead wood, leaves, and, mostly, spun silk. The species that lived or lives in this nest, *Camponotus senex,* is found throughout the moist Neotropics. The silk comes from the larvae—the same silk that is used to spin their cocoons as they become pupae. The workers carry the larvae around and hold them or put them down to spin the silk against the edge of a leaf or stem. In the coordinated fashion of ants, the bits of silk are neatly spun with small pieces of bark and dried leaves, plus particles of soil, which other ants bring up from the ground. It looks like a woven wall hanging using "found materials." Slipping halfway inside the silk nest, looking in, the bala ant has the sensation of being inside a cocoon—a large communal cocoon. She quickly retreats as the scent of the foreign ants inside repels her.

There are about a dozen species, four genera, of weaver ants worldwide—all within the advanced Formicinae subfamily of ants. The most advanced weaver genus, *Oecophylla,* found from Africa to Asia and Australia, dominates portions of the Old World tropics with its massive nests, which sometimes occupy a complex extending across several trees.

Suddenly a big weaver worker pokes her head out of another entrance to the nest. The weaver ant colony has smelled the heavy musky perfume of the big balas and is coming to investigate. The balas, no longer curious, race up the tree. As they move higher and higher, they begin to find different ants and beetles and other fauna in the canopy crown. It's another world, the most diverse part of the rain forest. It may be the most diverse ecosystem on Earth, and yet only very recently has it been thought of as an ecosystem separate from the tropical lowland or humid rain forest itself.

At ninety feet, just beneath the tree crown in the canopy, the balas smell something overpowering. The whole world is suffused with an intoxicating scent. At ninety-five feet they find blooming orchids on an entangled vine that almost blocks the route to the tree crown and the nectaries. Climbing around a flower four inches in diameter, the first sister locates the extrafloral nectaries on the red stripe of the stem, or petiole, of a leaf. She clamps her mandibles around the stem and bites into the plant's tissues until a droplet of juice rolls out. She lets it roll into her gaping mandibles. Then she turns around to her sister, touches her head with her antennae, and leans against her mandibles until they open, a reflex. She passes the droplet of nectar to her—a tender gift between the last two sisters of the bala colony. The sister tilts her head back and lets the nectar fall into her mouth—sweet refreshment, extra energy after the long climb. They then both grab hold of the stem and make new incisions, tapping more nectar droplets, which they each store in their social stomach.

A sweet tooth may seem odd in such a large, capable ant with big mandibles and a fearsome sting. As primitive hunters, however, balas rely on scavenging with only the odd kill. They are commonly seen gathering nectar or carrying droplets in their mandibles. It *is* odd but roughly equivalent to the discovery some decades ago that wolves in parts of the Canadian North subsist largely on mice, or that big African male lions lie around all day and leave the hunting to the females. For the ants, sugar provides quick energy.

Many ants use the sweet products available from plants as part of their diet. The source is sometimes the nectar found in flowers. Only a few ants, however, are active pollinators, and some plants produce repellents to discourage ants from visiting and wasting pollen that, from the plants' point of view, is better carried by flying bees and butterflies. Most ants make poor pollinators because they regularly return to the same branches or other sites. They also produce antibiotic secretions that may interfere with pollen germination. But many plants still want ants to visit, in effect preferring to work with them rather than against them. Such plants have evolved special extrafloral nectaries to attract ants or other helpful insects. These nectaries are sugar glands that have nothing to do with pollination and are located, depending on the species, on the stem or leaves or almost anywhere on the plant. The plant's strategy is to keep ants sweet. On the balsa tree, the extrafloral nectaries that attract bala and other ants produce a more concentrated nectar with a much lower amino acid content than the flowers that attract pollinating bats. To get their protein content, the visiting ants apparently have to kill insects that might bother the plant. According to ecologist Barbara L. Bentley from the State University of New York, Stony Brook, many plants turn on their extrafloral nectaries according to a schedule that attracts ants when the plant is most susceptible to damage, such as when the bud is opening or the seeds are being produced. The ants, by their very presence, tend to repel damaging herbivorous insects.

But some plants go to even greater lengths to provide more of an ant's needs

—in some cases, complete food requirements—so that the ants can move into a tree and never have to leave it. Besides extrafloral nectaries, many tropical plants offer special pearl bodies—tiny, single, or multicelled, oil- and protein-rich structures on the leaves or shoots. Certain acacia trees found in the dry tropics of Latin America and Africa have on the leaf tips bright orange pearl bodies, called Beltian bodies, for the acacia ants. And cecropia trees, such as the one that repelled the miniature army ants, have Müllerian bodies, numerous, small, glistening red, yellow, or white egg-shaped corpuscles buried in the dense mats of hair at the base of each leaf stalk. Beltian or Müllerian bodies are neat, brightly colored packages of food that are easily detached and carried around by hungry ants. In addition, both of these trees provide housing for ants—in the hollow thorns of the acacia tree and in the hollow trunk of the cecropia. The situation is so convenient that certain ant species have elected not just to visit regularly like the bala ants but to move in permanently.

FOR MORE THAN half a century before the 1960s there were two schools of thought about ant-plant associations. The exploitationist school, which included Wheeler at Harvard, argued that only the ants benefited and that the various plant structures that served as housing and food for the ants actually evolved for some other, unknown purpose. But Wheeler turned out to be mistaken on this matter, and his ecclesiastical nemesis, Father Erich Wasmann, turned out to be right. Wasmann championed the protectionist school, as W. L. Brown later called it.

The concept of ant plants originated with the same self-taught English ecologist who made some of the early observations of leafcutters—Thomas Belt. In his 1874 book *The Naturalist in Nicaragua,* Belt theorized that the ants were protecting acacia trees from competing plants and damaging insects because the acacias had evolved the pearl bodies, nectaries, and hollow thorns to feed and house ants. Belt is credited with the first description of this coevolutionary relationship, and for his discovery the ant food factories on plants still bear his name: Beltian bodies. But when Wheeler heard about this supposed relationship in the early 1900s, he and others challenged it, forming the exploitationist school and claiming instead that the ants were simply exploiting the situation as they found it. Father Wasmann promoted Belt's original idea—the protectionist school; however, neither Wheeler nor Wasmann ever devised a way to test the theory.

In the 1950s, Brown, fresh from his Australian travels, introduced new evidence in support of protectionism. Brown had observed in Australia that none of the numerous acacia species entered into associations with ants. Australian acacias had lost their spininess as well. Brown wrote that the absence of browsing mammals in Australia's recent geologic past explained the difference. He suggested that the presence of ant plants and spines—found in New World and African acacia species—was probably a response to large and effective browsing mammals. Certainly the acacia ants respond quickly and aggressively toward any size intruder. Brushing against an ant-occupied acacia tree, suggest Hölldobler and Wilson, is comparable to the sensation of walking into a large stinging nettle plant.

Brown pointed the way for tropical ecologist Daniel H. Janzen, then a graduate student at the University of California, Berkeley, later of the University of Pennsylvania. In a yearlong field experiment in the 1960s, Janzen finally proved that the association between bull's-horn acacia trees and the so-called acacia ants, *Pseudomyrmex ferruginea,* that lived in, on, and with them was symbiotic and mutualistic—both species benefited from the deal. He did the first experiment in Mexico, then continued working with the same and similar ant plants in Costa Rica. First, he removed the ants from a number of acacia trees, in some cases even clipping the thorns or spraying with an insecticide called parathion. Right away these acacias, ranging in size from shrubs to small trees, were attacked by insect herbivores, including various beetles and caterpillars. Other plants and trees began to crowd in close and shade the affected acacias; they began to look as vulnerable as nearby acacias that had never had any ants on them. Meantime, Janzen's control trees, which he left the ants to defend during the year, kept up an impressive performance, with a quarter of a colony patrolling the leaves night and day, cleaning them and almost always successfully killing or driving off invading insects. The ants also chewed to death any sprouts from alien plants within a radius of sixteen inches around the trunk and destroyed anything that touched the tree's upper branches. By the end of the year, the trees that had had their ant colonies removed were in steep decline. Some were not even going to be able to produce seeds before they died.

The story of the ant and the acacia tree made a deep and lasting impression on Janzen. It became a sort of parable for the complexity of the workings of insects and plants in the tropical forest. It taught him patience and encouraged him to move permanently to the tropics to become a dedicated field biologist and conservationist. He wanted to know everything he could about everything —the millions of stories writ on and underneath every leaf of the tropical forests.

Besides the fierce acacia ants on the acacia trees and the aztec ants on cecropia trees, there are some ant-plant associations that are much less fierce, yet just as intimate and effective. The tiny *Pheidole bicornis* ant has long been known to live in *Piper* plants in Costa Rica, but entomologists have pointed out how inept these ants must be as defenders. They were never taken seriously as ants that could repel a plant's invaders. But every ant has its champion. University of California, Santa Cruz, myrmecologist Deborah K. Letourneau, who has spent several years studying this ant-plant association, found a subtler side to competition. Instead of fighting other adult ants and insects, these *Pheidole* ants are methodical at removing the eggs and larvae of plant-eating insects. Her work has shown that *Piper* plants inhabited by this ant, found at La Selva and elsewhere in Costa Rica, suffer much less damage than those with no ants. In particular, the workers patrol the susceptible new leaves, keeping them free of anything that might later grow into a problem. Effective competition is not always flashing swords and hyped-up aggression. Having a large, aggressive workforce can be a real drain on the food resources of a colony. Compared to the aggressive acacia ants and the aztecs, the *Pheidole* ants protect their plants with much less effort. In exchange, over the period of a little more than a month, the ants' presence stimulates the production of pearl bodies. Plants without ants might have fifteen to twenty pearl bodies per leaf, but plants with ants produce about fifteen hundred.

The ant-acacia tree story remains one of the strongest and best studied examples of the coevolution of ants and plants. Such coevolution dates from the mid-Cretaceous period when flowering plants came on the scene shortly after ants and other social insects. While birds and bees became the great pollinators and have taken the leading role in coevolution with plants, the ants, too, have played a part. At least sixteen species of plants at La Selva are specially adapted for ant occupancy. The story of the cecropia tree with the resident aztec ants is not as clear-cut a case of coevolution as some, in part because experiments have revealed that removing the ants does not always result in large-scale insect attacks, and some cecropia trees apparently survive with no ant colonies aboard. But some researchers suggest that the cecropias' lack of built-in defenses, such as are found in many other rain forest plants, may have led them to their relationship with the aztecs, which have in effect become the tree's chemical defense system. Still, other cecropias have resident aztec ants that are less aggressive or at least less thorough at keeping other plants and insects away from the tree. Many of these, however, are young aztec superorganisms. The bigger the colonies get, the fiercer they become and the more they dominate the tree. And this association—like the ant-acacia tree story—is longstanding, millions of years old. Aztec ants have been found in amber deposits from the Dominican Republic dated some 20 million years ago. The aztecs look much as they do today and were probably living in cecropia trees even back then.

THE BULLET SISTERS, satiated for the moment, traverse the high branches of the tree. They soon approach a lush garden of epiphytic plants nestled snugly against the trunk and a main branch. Epiphytes are plants of various species that live on other plants. It is a lifestyle classification; many plants, such as various orchids and bromeliads, can be epiphytes. The common strategy of an epiphyte is to evolve novel ways of keeping and collecting water yet obtaining sunlight crucial to photosynthesis. Life in the dark understory of the rain forest forces them to move high above the ground, closer to the light source. Yet they must develop special leaves that hold or absorb rainwater and, in some cases, giant cisterns that collect water and nutrients as they provide homes for aquatic insect communities. Others send out aerial roots for Tarzan, which are actually growing in search of water that they can suck up to the plant. In some cases these roots steal water from the basins of other plants or from a crotch of the tree where water collects.

Approaching closer, the balas can smell various plants—figs, climbing vines, and bromeliads. It is an ant garden—another mutualistic symbiosis between ants and plants. Ant gardens, nearly spherical or ellipsoidal in shape and up to two feet in diameter, exhibit in usefulness what they lack as classic floral arrangement. And coming out to greet the bala sisters are the acrobat ants, *Crematogaster longispina,* the ants that plant and tend these "ant gardens." Acrobat ants are easy to recognize from their characteristic heart-shaped abdomens, flat on top, convex below. Their gardens begin when worker ants gather small fruits and seeds and plant them in the walls of their aerial nests. According to Jack Longino, who has studied and collected many La Selva ants, the acrobat ant colony manufactures a loose, coarse-fibered material for its nest, from plant fibers, detritus, and bits of soil, and this so-called carton provides a physical substrate as well as nutrients in which epiphytes can grow. As each wall, roof, or gallery is laid down, a new crop of epiphytes sprouts from the carton, making a dense tangle of roots, stems, and leaves.

Carton building—different from weaving, which uses silk threads—is widespread among the ants, but the carton builders that cultivate epiphyte gardens have evolved some peculiar characteristics. The ants are strongly attracted to the fruits and seeds of their garden plants. Myrmecologist Diane W. Davidson from the University of Utah has suggested that the seeds of the particular species they collect contain pheromone-mimicking chemicals as well as food. Even when the seeds sometimes pass through the digestive tracts of bats, they still remain attractive to the ants. After the ants consume the elaiosome, the aril or fleshy covering of the seed, and any traces of fruit pulp sticking to it, the ants plant the seed.

The gardens of the acrobat ants form one of the dominant features of La Selva and other lowland tropical humid rain forests of Central America. Ant gardens occur almost exclusively among several ant species in the Neotropics, including a carpenter ant, an aztec ant, and a thief ant *(Solenopsis),* though a few more ant gardens have recently been discovered in tropical Asia. "Ant gardens" should not be confused with fungus gardens. The ants that tend ant gardens are completely different from the leafcutters, who belong to the so-called gardening or fungus tribe of ants, the Attini. The leafcutters and other attines live in the ground where they grow fungi. The ant gardeners, the species that tend aerial gardens, are actually various unrelated species that live in trees.

The ant gardeners may have coevolved with the plants in their gardens, but this is not clear. At least one plant species, a vine called *Codonanthe crassifolia,* grows faster in the ant gardens than on its own, and the seeds won't germinate at all on the ground or in any medium but detritus or ant carton. As for the ants, they collect the vine fruits and eat the pulp; the extrafloral nectar is used to make the carton, while the roots of the growing vine provide structural support to the nest. But the ants do not help pollinate the plants or guard the extrafloral nectaries of the mature plants. The ants also forage for other food outside the nest. Even though the ants do not rely entirely on these plants, and vice versa, the ant gardens show evidence of some symbiosis.

Inside the acrobat ant garden is another surprise, for these are one of the many ant species that keep "cattle." Deep inside their carton galleries they have separate shelters for various species of homopterans. The order of insects called Homoptera includes aphids, jumping plant lice, treehoppers, froghoppers or spittlebugs, whiteflies, coccids or scale insects, mealybugs, and related groups. Rather than attack and kill these soft-bodied, relatively defenseless insects, ant colonies have evolved a clever, cooperative relationship that resembles a dairy farm.

The relationship often starts when ants bring aphids or other homopteran eggs into their nest. In other cases, adult homopterans approach the ant nests and, after exchanging licks to prove how attractive they are to an ant colony, are invited to stay. The aphids, scale insects, and other "cattle" have stylets, sucking mouth parts, which are specially designed to pierce the hard cuticle of plants and to feed within the cell walls. The aphids often puncture the vital, main transport stream, or phloem vessels, of the plant. This main transport system is under pressure, which simplifies the aphid's job once the system has been tapped.

The homopteran digestive system is specially adapted to handle this liquid plant diet. As the aphid or homopteran takes its nourishment, it produces water and carbohydrate by-products, which are eliminated as liquid feces known as honeydew. The honeydew "waste" flows fast and thick in these insects, and the quantities are substantial enough to interest ants and many other insects as a valuable food. Besides water, sugars, and other substances, honeydew contains key proteins, free amino acids, minerals, and vitamins. The honeydew is eagerly sought by some other social insects, too, including certain bees and wasps, as well as various flies and other organisms. It has also been used for human food when it is sometimes secreted by masses of insects and collects in large quantities on the ground. Honeydew is thought to have been the miraculous manna from heaven, celebrated in the Old Testament—the food that saved the Israelites. Modern Arabs still collect and use these secretions which they call "man."

Yet of all the honeydew collectors, including humans, only ants actually "milk" the honeydew, collecting it directly from the aphid or other homopteran. The exact procedure varies from species to species, but at intervals of only a few minutes, an aphid will raise its abdomen, letting the ants know it is ready to be "milked." An ant approaches and licks the abdomen. Sometimes, to enhance these visual and chemical messages, or in case they don't get through, the ant also drums with its antennae on the aphid's abdomen. Seconds later the droplet of honeydew appears, which is then taken by the ant in its mandibles and swallowed. The chemical communication between the ant and the homopteran may go something like this:

ANT: Okay, we recognize you, aphid. How about some food?
APHID: I read you, ant. I'm waterlogged and need some relief now. Take it, please.

One worker ant can tend several homopterans by steadily moving from one to the other, approaching each in turn from behind and licking the secretions. The whole effort is a little like milking a barn full of cows that recently gave birth and need to be milked every few hours. The liquid is taken into the ant's social crop and held for later distribution to the rest of the colony.

Sometimes the ants kill some of the aphids or other homopterans, particularly when there are too many of them or when extra food is needed to feed their own growing larvae. Yet the ants always maintain healthy-sized herds to keep a steady flow of honeydew. It's easy to see why the aphids and other homopterans allow themselves to become cattle when the ants conveniently relieve them of their honeydew, but why do they stay if the ants kill them when they don't need them?

Removing the honeydew means the aphids attract less predation and contamination from fungal disease. In many cases, the ants build special galleries or devote areas of their nest to their cattle, sheltering and protecting them, and providing plant material to eat. But even in the absence of shelters, ants often protect the homopterans from predators and parasitoids. Life without ants can be much worse: Certain flies, beetles, and other insects take honeydew from homopterans with nothing offered in return. In sum, for aphids and other homopterans, survival seems to be much better living as guests in ant colonies than living outside and away from them.

Besides residing in the gardens of the acrobat ant, homopteran cattle live among many other ant species. The aztec ants keep mealybugs mainly for quick-sugar fixes, but the mealybug cattle may be crucial for the success of aztec colonies. Aztecs *with* cattle can afford to be hyperaggressive, keeping all other competing ants away from their precious Müllerian bodies. The aztecs with the most cattle may well have the best chance of survival.

A rapidly growing population of aphids or other homopterans could inflict serious damage on, if not devour, a plant or tree. The plants in the ant gardens and the stems of cecropia trees seem able to tolerate some damage from aphids and other homopterans, especially if they are the guests of ants. Ants, in effect, handle the pest control. And as anyone familiar with pest control knows, the goal is not elimination of a pest but management at a tolerable level. With ants in charge, damage from the aphids or other homopterans is kept at a tolerable level.

Certain homopteran species develop part-time relationships with ants, yet retain their ability to live without ants. Many other homopterans, however, have coevolved with ants. They are truly domesticated as cattle and unable to live except as guests of their keeper ants. These myrmecophiles—from the Greek words meaning literally "ant lovers"—may be associated with various species of ants. Myrmecophiles have anatomical peculiarities and behavioral adaptations, such as special bristles around their anus that hold the honeydew until the ant removes it. Part-timers or anti-ant homopterans have no bristles and, when unattended by ants, dispose of honeydew droplets by flicking them away with their hind legs. But some aphid cattle, like highly bred domestic animals, have lost their protective structures, and others can no longer use their legs for jumping.

The parallels to cattle ranching are apt, albeit with a few bizarre twists. Just as humans keep cattle to turn grasses and grains into milk and meat, ants have domesticated homopterans to utilize plants. Some ant species direct their cattle to the best parts of the plant for feeding. Other ants invite their cattle inside the nest, sometimes keeping them in separate chambers, protecting them from parasites and predators, and moving them with their own brood if they need to relocate during the lifetime of the colony. A few ant species, including *Acropyga,* which lives in the ground at La Selva, even carry their mealybug cattle in their mandibles during the nuptial flight. Once they've mated, this "dowry" from their previous colony ensures a ready food supply.

Still another ant species has taken on the role of nomadic cattle herders—the nomad ant, *Hypoclinea cuspidatus.* These ants move their mealybug cattle herds and the colony as needed. They are the only ants that are nomads and arguably the only truly nomadic animals, according to a strict definition of the word. Living deep in the Malaysian rain forest, they have entered into a long-time arrangement with some of the local mealybugs. A mature colony of nomad ants consists of perhaps ten thousand workers clinging together to protect a queen, some four thousand brood, plus more than five thousand mealybugs. The mealybugs, which associate only with this ant, feed on the phloem sap of the new parts of various flowering plants. As feeding sites are exhausted, the ants transport the mealybugs along an intricate trail system to new pastures, some more than sixty-five feet away. When the food is much farther away than that, however, the ants simply transport the colony nearer the food source. And these ants and their mealybugs are totally dependent on each other. When German entomologists Ulrich Maschwitz and Heinz Hänel separated them, the ants declined rapidly, while most of the mealybugs walked around aimlessly and began to die from the contamination of their own honeydew, which was not being removed. When other ant species visited the mealybugs, the mealybugs were not accepted but instead were attacked and killed as prey.

Ant colonies maintain cattle around the world, from temperate through tropical regions. The practice occurs in most of the members of the three most advanced ant subfamilies, the Myrmicinae, Formicinae, and Dolichoderinae. Even some of the ponerine ants, known for their hunting prowess, occasionally take nourishment from other insects. The acacia tree provides all the food needs to the acacia ants, but the aztec ants tend herds of mealybugs inside the hollow cecropia tree trunk, nurturing and protecting them, to supplement what they harvest from the tree's pearl bodies.

But the bullet sisters are strangers to farm life and such domestic customs. The balas, save for their craving for sweet extrafloral nectaries, are primitive hunters and scavengers. They often see the cattle kept by other ants, and watch the honeydew milking, but it is another world, as different for them as life on Mars or another planet would be for us.

• • •

NIGHT IS PASSING. The bullet sisters begin to walk back down the tree, retracing the original trail laid by the one bala. But where are they going? What do they do without a colony? All that seems to be left to them now is to keep moving, eating when hungry, staying alive. If there are other programs left in their genes, things that might change what is left of their thin lives, they cannot sense it. For now, all is survival.

Down on the ground, the mini army ants have been scavenging and killing all night and now have enough food to satisfy the hunger of the eighty thousand ant colony. At least for another day. The catch for the night includes more than seven thousand various ants, mostly smaller workers from the many colonies found in the rain forest leaf litter. About a quarter were already dead, the carcasses first come, first served. The others were alive and required some combat and overpowering by the wriggling, stinging arm of the mini army ants. The colony also collected some sixty-seven beetles and assorted beetle parts, eleven flies, and miscellaneous springtails and other invertebrates. It is a massive catch for a single ant superorganism, and it shows how effective miniature army ants, all working together, can be at killing and retrieving prey.

A few of these ant kills came early in the night when the mini army ants crossed a leafcutter ant path and cut off a group of workers dismantling a young tree that the leafcutter scout had located. After retreating from the first waves of army ants, the leafcutter scout had ventured out again a couple of hours later, found a productive area near the edge of a natural clearing where two trees had fallen the previous year, and directed several lines of foragers there. Twenty minutes into their labors, the ambush came. Before the leafcutter soldiers could be mobilized, the army ants surrounded several leafcutters in a ratio of ten to one. Only six leafcutter workers died, all still carrying their leaves—as if refusing to surrender them or using them as shields. But once intercepted, many simply had no time to drop their leaves and fight or flee. Meeting the full force of an attacking column at the peak of its midnight siege, they had no choice but to suffer the losses and carry on.

But now the miniature army ants are deep into their retreat as morning approaches. The typical pattern for most army ant species is four to five hours of foraging followed by a midday siesta or period of low activity, and then a general retreat in which the tired workers return, some carrying meat back to the booty caches and to other members of the colony around the bivouac. During the return, which sometimes lasts for hours, there is some catching and killing, but little compared to the early forays.

As the retreating mini army ants file past, avoiding the main leafcutter highway, the leafcutter scout waves her antennae from side to side and tilts her head to the sky, looking for any sign of light. The leafcutters' work is at its peak now, and the mini army ants will no longer interfere. But the witching hour of dawn, when the swarm raiders, those giant army ants, break camp and fan out, is no time to be out on the open road. There is no sign of morning yet. The main highway is dense with leafcutter workers coming and going. The scout steps off the highway and into the leaf litter, where all the foreign smells begin. The scout will try to make another foray before dawn in search of new leaf material. It may not be useful for tonight's harvest, but it would be ready for tomorrow. Stepping quickly through the crackling leaves, the scout advances to a new area beyond the clearing. She finds a number of promising young seedlings, but nothing worth dismantling yet.

Just as the light of day begins to leak into the rain forest, her routine exploring is disturbed by a chance encounter. On the way back, taking a detour through a slightly damp portion of the forest where the colony seldom visits —in part because it floods most years—she encounters a bizarre sight. She smells ant death and then, in the dim light, she glimpses one of the carpenter ants, *Camponotus*—a successful genus that has radiated widely, with many species living around the world. This carpenter ant, however, is not so successful.

At first the leafcutter ant almost recoils at the grotesque sight. The bizarre ant is standing at attention, frozen to the spot, with a stalk almost four inches high protruding from the back of its neck. On closer inspection, the ant is clearly dead and reeks of the parasitic fungi that has infected the corpse, taking over its entire body, almost masking the necrophoric substances that ants release upon death. Called *Cordyceps,* this giant, brightly colored fungus invades and kills several species of ants. The scout, perhaps coincidentally, turns her head and cleans her right antenna by sliding it against the comblike tibial spur on her right foreleg, then turns the other way to clean the other side. Next she leans over and cleans both forelegs thoroughly with her tongue. Finally, bending her body almost in half, she reaches back to the tip of her gaster and gives that a thorough licking as well. She collects this dirt into a ball and presses it into the infrabuccal chamber, the small pouch located on the floor of her mouth. Other workers back in the nest would do the cleaning for her, but out here on the trail she has to do it herself.

Ants are very clean animals. The wide trails of the leafcutter ants, in particular, are nearly spotless roadways, and the cleanliness is most crucial underground. For a creature that spends so much time in warm underground galleries and in rotting wood, living so close together, it is astonishing that fungi do not take over whole ant colonies. Of course, the leafcutter ants grow and harvest their particular fungi gardens, keeping them under tight control, harvesting them for food, and never giving them a chance to produce spores. They also weed out any competing fungus species. There are many other kinds of fungi and other microorganisms that would seem to have perfect conditions in ant societies for luxuriant and unfettered growth. A few of these fungi live in or on ants. But fungi on ants are comparatively rare—even in the wet lowland tropics. The reason is that workers frequently clean themselves and their companions with their forelegs and tongues. The ants periodically cough up the infrabuccal pellets, the waste materials, sometimes inside the colony chambers, where it is promptly removed as waste, and sometimes directly into special refuse pits in or near the nest. As part of their cleanliness habit, ants also secrete a number of potent antibiotics principally from the metapleural gland, near the rear of the midportion of their body. The main substance in these antibiotics is phenylacetic acid, which fights fungi and bacteria. Among all the social insects of the order Hymenoptera (the bees, wasps, and ants), ants are the only ones with the metapleural gland. Bees and wasps protect their broods in specially constructed, antibiotic-impregnated brood cells, but the ants have their larvae and pupae in open chambers or tunnels. It is the metapleural gland that allows ants to colonize the ground. Their industrious cleaning keeps their broods and queen from parasitic invaders. The few ant species without a metapleural gland are those that live in the trees—much drier and cleaner environments. For them, regular grooming seems to be enough. Monkeys are not the only social animals that spend time sitting around grooming each other in the trees.

The metapleural gland—and the seeming neurotic cleanliness of ants—is part of the reason for the ants' worldwide success. This and the ants' social habits—the capabilities of the superorganism—enabled ants to colonize the Earth with every conceivable adaptation and in great numbers. This is the triumph of the superorganism.

THE SUPERORGANISM CONCEPT seemed brilliant, ambitious, and inspirational for several decades after Wheeler originated it. By the early 1960s, however, the expression was no longer used in scientific circles. The reason, according to Wilson, writing in 1971, was "not because it was wrong but because it no longer seemed relevant. . . . The concept offer[ed] no techniques, measurements, or even definitions by which the intricate phenomena in genetics, behavior, and physiology [could] be unraveled."

These were the words of a myrmecologist working hard, along with his contemporaries, to study all the minute parts of the colony, to look at each aspect of colony life using experimental and other techniques. With that effort yielding substantial results by the 1980s and 1990s, Wilson and Bert Hölldobler revised their view of the superorganism: "The time may have arrived for a revival of the superorganism concept."

The past several decades of studying ant societies has produced a new understanding of the ways in which castes are determined and their actions coordinated through various kinds of communication. Wilson suggests that a next step may be to compare the development and organizational processes of societies to the growth and differentiation of tissues in individual organisms. On the horizon they see a future general theory of biological organization. And so the old superorganism concept now thrives again as part of the dreams of modern scientists to come up with the Big Biological Picture.

As the leafcutter worker walks home to her nest, she passes colonies of ants at every step—each a successful superorganism in its own right. The competitive edge comes from the organization of an ant colony with its various castes, each performing a different function. Like a supercomputer capable of "parallel processing"—doing many tasks at once—some worker ants expand the nest, search for food, feed the larvae, feed the queen. Others remove the brood—the eggs, larvae, and pupae—to a safe place during an attack. Large workers, or soldiers in many species, use their oversized, sharp-toothed mandibles to defend the colony.

Such specialists make fewer mistakes, and the work of each task goes faster. At the same time, the food supply is stabilized as food reserves are stored in alimentary, or food, eggs laid by the queen and some of her workers. Because groups can focus all their force and energy, it can be harnessed to accomplish such monumental projects, in the case of the leafcutter ants, as setting up vast underground gardens. Perhaps the Portuguese explorers who went to Brazil and encountered the leafcutter ants and other insects in the sixteenth century said it best. After seeing the great productions of the ants and noting their extraordinary numbers in the vast Amazon, which scientists have more recently calculated at 3 million ants per acre, the Portuguese reported back to Europe: "Ants are the king of Brazil." In fact, like the notorious Amazons—the tribe of human female warriors from Greek mythology—ants are largely female societies built around one or more queens. The Portuguese should have called ants the queen of Brazil, for theirs is the queendom of the earth.

To an astonishing degree, the *successful* insect faunae of the world, such as ants, are social. Sociality confers certain advantages on any animal, but with insects it can be said to equal success. The social ants, bees, and wasps, together with the termites, represent an estimated 75 percent of the total insect biomass, the weight of all the individuals taken together. One-third of the entire terrestrial animal biomass in the Amazon is composed of ants and termites. Worldwide, an estimated 10 percent of the animal biomass is ants.

Success can also be measured in numbers of individuals. With insects, because they individually weigh very little, a high biomass weight means an extraordinary number of individuals. One conservative calculation, made by E. O. Wilson, suggests that ten thousand trillion ants may be alive at any one time. He loves to cite a litany of examples to show the importance of ants; among them: In the Ivory Coast savanna, ants have densities of more than twenty-eight hundred colonies and 8 million individuals per acre. In parts of southern Africa, the "driver" army ants contain densities as high as 20 million workers in a single colony. But the ultimate ant social system is that of the highly evolved *Formica yessensis* ant on the Ishikari coast of Hokkaido, Japan. One multiqueen supercolony contained 306 million workers and some 1.08 million queens. They lived in forty-five thousand interconnected nests stretched across a square mile. The density per acre was about seventy nests containing some half a million workers. Wilson talks about the little things that run the world, by which, of course, he means ants.

The great seeming paradox with ants is that these numerous insects can also be mainly carnivorous. In ecology, the large carnivores such as killer whales and wolves have populations that number in the hundreds, compared to prey, which number in the hundreds of thousands. Ants would seem to break that rule. How is this possible? If ants are considered *only* as superorganisms, then their numbers are much, much less. The mature leafcutter colony with 3 million to 7 million workers is reduced to one, as is the army ant colony of close to a million. But many, if not most, ant colonies are unassuming productions of less than a hundred individuals. The average for all ants might be on the order of one thousand individuals per colony. Thus there may be 10 trillion ant superorganisms—still a high number. But there are other explanations for how ants can be numerous yet dominant. First, some of the other insects that ants largely subsist on may have even more astronomical numbers than ants. We are just beginning to realize how diverse and numerous many insect groups are. It is entirely possible that the pyramid that includes the ants' insect prey is much larger than we know. Second, ants feed extensively on other ants; therefore, part of the large ant numbers are prey as well as predators.

How ants can be dominant atop the food chain and so numerous probably depends more than anything else on their intimate relationships with plants. Ants are not only carnivores but also herbivores. For the most part, ants don't eat plants directly as do grazing mammals or many other insects. Ants are mainly carnivores, and plants are unpalatable to them. Instead, they have evolved ways of coping with the defensive chemicals and unpalatable materials of plants and turning them into something edible. Leafcutter ants get around the problem one way—by using a fungus to act as a kind of digestive system to break down the leaves and transfer the nutrients into edible form. In this system they use an enormous amount of plant material. Extrafloral nectaries and pearl bodies allow other ants limited access to the food products of certain plants—sometimes in exchange for the ant performing protective services. And a third main method is to enlist herds of herbivorous aphids and other homopteran insect "cattle" to predigest the plants and turn them into a usable form: honeydew. Turning grasses and leaves into milk and meat is precisely what numerous species of ant cattle do. Myrmecologists describe this relationship as so efficient that there is almost no energy loss in the honeydew transfer. This means that ants are living essentially as primary consumers, as well as carnivores high in the food chain.

MEANTIME, DOWN ON the ground, the leafcutter ant scout has finished her work for the night and heads back. The dawn bird chorus, featuring the whooping oropendolas, starts up and gains steady support for the idea of a new day. A few miles downstream along the Río Sarapiquí, a howler monkey bellows to mark the start of his morning rituals. The early-morning light glints off the giant carton nests of the weavers up above in the canopy. The mini army ants are now a few stragglers plus the inevitable "ecitophiles"—the insects, termed army ant lovers, that follow the army ant colonies, picking up the scraps.

But something is stirring: a massive beast, the largest, most powerful animal in the American tropics—the swarm raiders, *Eciton burchelli.* It is one of the ultimate insect superorganisms. The sun picks up the color in the big, hard, mostly dark bodies and turns them into a flickering reddish orange. Their long, sturdy legs make them cast a long shadow. At first, the charge follows a treelike pattern, columns of ants gradually splitting into branches, similar to the long columns of other army ants. On average, individual workers are two or more times the length of a miniature army ant. The overall colonies are also much larger. The foraging arm of a swarm raider colony contains up to 200,000 ants. At up to about eighty feet wide, it is more of a slow, monstrous steamroller than a foraging "arm" as with the other species of army ants. In all, the column raider colony *Eciton hamatum* has 50,000 to 250,000 workers and the miniature army ants of the tropics 10,000 to 140,000 workers, but a swarm raider colony can have 150,000 to 700,000 members.

About ninety minutes after dawn, the swarm raiders begin fanning out as they go for the kill. The wide swarms seem to be driven by the fan-shaped network of columns as they pour into the front like a river. They are moving at a slow but steady fifty feet per hour. They don't want to miss anything.

The jungle seems momentarily mesmerized, with frogs, lizards, and small mammals looking this way and that, and insects trying to catch a whiff on their antennae. But they are all soon thrown into panic. As the swarm builds, insects flee all around—from armored beetles to tiny mites—some flying, others running or burrowing. Hundreds of ant colonies also flee, the workers carrying the brood and sometimes the queen. Others whose strategy is to try not to be noticed freeze on the spot. Still others go for the bunkers of their own colonies, blocking entrances, hoping they won't be noticed. Even big-eyed lizards, snakes, tarantulas, and scorpions have to break camp or die before high noon. The swarm raider superorganism, as it builds to its widest and most powerful phalanx by mid-morning, will overwhelm almost anything in its path—even nestling birds and small, disabled mammals—carrying the booty back to the bivouac where the hungry larvae wait.

Unlike the nests of the leafcutters, bala, and most other ants, which are situated underground or in pieces of wood, the swarm raiders make a portable nest out in the open, largely from their own workers. The workers entwine their long legs to form curtains of ants hanging from logs or branches, and inside, the queen and the colony larvae and pupae are kept warm and safe. Most nineteenth-century naturalists to the Neotropics, such as Thomas Belt in Nicaragua, could not find the hive or colony home of the army ants, despite trailing them sometimes for half a mile or more through the jungle. They didn't know what to look for. American naturalist Edward Norton, working in Mexico in 1868, finally found army ants assembled "under a fallen trunk, a prodigious number of workers . . . heaped and piled upon each other like the bees in a swarm."

There are several adaptive advantages to the army ant bivouac. Because it's entirely portable, the army ants can move to their food. By crowding closer together or moving apart, they can regulate their own temperature. Also, they can expand the size of the bivouac as the broods and colony increase, encouraging the pragmatic growth of the superorganism. A column raider bivouac can accommodate eighty thousand at a time, but a swarm raider bivouac may contain three hundred thousand.

In the pink light of early morning, the bullet sisters look on from a branch a few feet above the seething mass. They can see all the way back to the swarm raiders' bivouac, more than a hundred yards down the trail. After watching for a few seconds, one bullet sister steps aside and lurches into a sticky patch of tree sap. First she finds that her front legs are caught. Then, as she instinctively touches her antennae to the amberlike resin and tries to bite it with her mandibles, her whole head becomes engulfed. Like quicksand, it begins to devour her alive. It is a slow death. As if noticing nothing, the last bala sister simply keeps walking up the tree until she picks up her trail to the extrafloral nectary. Swinging her head periodically from side to side, she waves her antennae to catch the molecules. She is not hungry and her superorganism is long gone, yet she keeps climbing. Reaching the extrafloral nectary, she instinctively picks up a droplet, a bead of sweetness, holding it in her mandibles. And then she heads back down the tree to her empty nest, looking for a sister, any ant, to share her find with. She carries the sweet bead high as if she had all the purpose in the world.

Chapter Four

Amber Tales

Ants entombed, preserved forever in amber,
a more than royal tomb.

—Francis Bacon,
HISTORIA VITAE ET MORTIS, *1623*

The sense that insects belong to a different world than ours is shared by many people, and it is a perfectly valid feeling. After all, the search for a common ancestor of insects and ourselves would take us back more than half a billion years. [But] insects are very much of this *world, and* Homo sapiens *is a strange and aberrant creature of recent origin who has sought to create his own world, apart from that of nature.*

—Howard Ensign Evans,
THE PLEASURES OF ENTOMOLOGY, *1985*

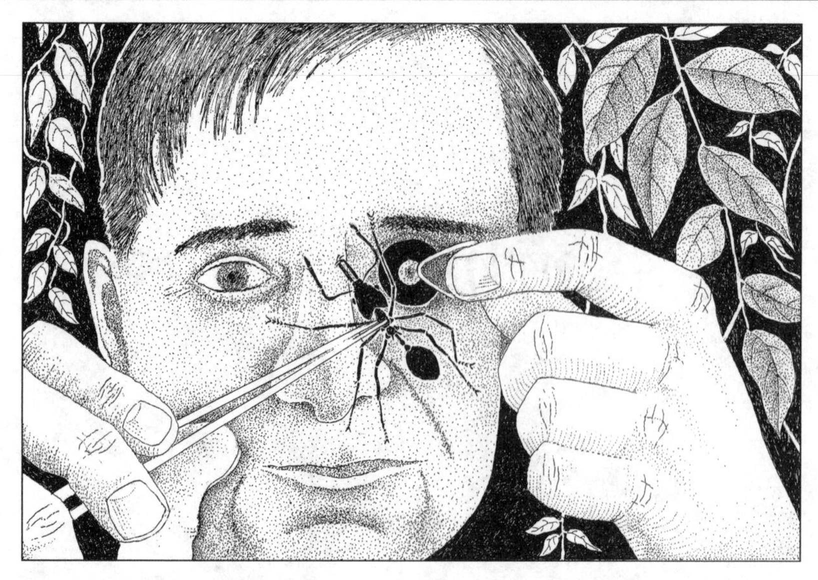

Wilson's Dacetines. E. O. Wilson holds up one of the elegant dacetine ants for a closer look with his jeweler's loupe.

PERIODICALLY through the morning—almost paralleling the path of the army ants, or a little in front of the swarm—Wilson has been squeezing sticky stuff onto tree trunks and branches, taking note of the date and location. The gray opaque goop is something called Tangle Trap, made by the Tanglefoot Company of Grand Rapids, Michigan. Wilson and Brown call it just "Tanglefoot." It comes in a tube similar to toothpaste. The label describes it as "an insect trap and adhesive used to repair sticky traps to catch and hold insects; for use in orchards, farms, gardens and homes; weatherproof and long lasting." It is also "sometimes used to collect data in support of ecological and control studies."

Wilson is particularly interested in it because it has similar properties to amber—those hardened lumps of fossil plant resin in which ancient insects are sometimes found entrapped. Amber, with its yellowish to orangish brown glow, is popularly used for jewelry, but the amber that has managed to ensnare insects allows entomologists and other paleobiologists to piece together details of the evolution of ants and other insects. That's because, amber comes from various identifiable parts of the world as well as from identifiable periods of geological history. Amber can tell you about the ancient insect faunas of specific places such as the West Indies, northern

Europe, western Canada, or New Jersey. It can tell you which ants walked the Earth, from the mid-Cretaceous period 80 million years ago through most of the Tertiary period to the early Miocene epoch, approximately twenty million years ago. By comparing this information with what is known about current ant species, you can get a hint of how certain species may have evolved.

Besides the search for oddities such as the miracle ant, with its pitchfork-shaped mandibles, Wilson's task for this expedition is a pilot study to see which insects might become entrapped in amber if it instantly created fossils of modern fauna. As part of their passion for ants, Wilson and Brown have coauthored several of the key papers on ant fossils and evolution. Using Tanglefoot, Wilson hopes to make a careful comparison of what was caught in amber back then with what might be caught in similar circumstances today in the Central American rain forests. This might help provide a clearer understanding of which kinds of ants and other insects get trapped in amber and under what circumstances, and how representative the amber fossils are of the time when ants first began to walk through the forests and deserts, forming into the complex societies that have long dominated the Earth.

As Wilson wanders from tree to tree, squirting out Tanglefoot and making notes in his notebook, he and Brown launch into an argument on the value of fossils in determining the evolutionary histories, the phylogenies, of the ants. There are two main ways to reconstruct phylogenies. The first is through arranging living species into an increasingly inclusive evolutionary tree based on the number of features they share in common. The other is to look at fossils with the hope of finding "missing links."

With insects there has been much debate about the value of fossils. Most fossils tell us nothing new about evolution because they fit neatly into preexisting groups. The fossil record for insects remains incomplete. The chitinous bodies of insects do not fossilize nearly as well as the calcareous shells of mollusks or the bones of vertebrates. Amber is the main fossilizing material for insects, yet most of it contains ants that are more or less the same as those living today. With a few notable exceptions, the existing amber fossils offer either nothing new, or tantalizing glimpses or fragments of ancient insects that fuel further argument.

Yet both Wilson and Brown have long been believers in the potential value of amber fossils—Wilson more passionately than Brown. Wilson is perhaps a little more of the prospector who likes to search for missing links and craves the thrill of finding one. Wilson has an arrangement with several commercial collectors, who purchase and send to Harvard's Museum of Comparative Zoology all the amber they uncover from productive sites such as the Dominican Republic. In an effort to deal with the steady stream of material, Wilson often carts the rough amber to his home laboratory for grinding and polishing and a first look under the microscope. According to Wilson, "the Chinese"—who have some amber ants that Wilson wants to study—"have a saying, 'One peek is worth a thousand finesses.' Well, I'd like one peek at those Chinese fossils!"

Wilson declares that fossils "sure beat the painful, painstaking reconstructions based entirely on the fauna as it is today." Brown, who has done some of those painstaking reconstructions, doesn't need to be defensive with Wilson, and isn't, but he reminds Wilson about how frustrating some of the amber fossils can be: "When you do finally find something that might be valuable, you get the body folded over, obscuring the petiole or some other vital characteristic you so desperately want to see." But to Wilson it's all a great jigsaw set in space and time, with a few pieces beginning to provide details of the emergence of a major group of animals, the ants, and he feels "sheer delight" to be a player, to be able to put some of the major pieces in place on the board, however tentative some of them may be.

"We need both methods to study insect evolution," says Wilson, and Brown concurs. But Wilson also sees the value of fossils for his studies of biogeography. "No one could have guessed some of the things we found when we first looked at the Dominican amber. Some of the species that were around twenty million years ago in the area that became the Dominican Republic are gone today, and others have moved in to take their place. You'll never get that kind of information from just looking at modern-day fauna." Wilson has also discovered and named three extinct species in the Dominican amber.

But the main jigsaw piece, or missing link, thus far that amber has provided is the "wasp ant," *Sphecomyrma freyi*, the first Cretaceous ant remains. Wilson, Brown, and the late Frank M. Carpenter, Harvard's longtime fossil insect authority, studied and reported on these remains in 1967. The fossil came from a retired couple living in suburban New Jersey, hobbyist rock collectors who found the amber piece at the base of some seaside bluffs at Cliffwood Beach, a few miles south of Newark. It contained two worker ants in a single piece of amber. The fossil was taken to nearby Princeton then to Carpenter at Harvard, who offered it to Wilson, who promptly proceeded to drop it on the floor, splitting it in two. Conveniently, each piece now held a single worker, fortunately undamaged. Wilson polished them to reveal stunning views of two ancient, primitive ants that had walked the Earth with the dinosaurs. They had very short wasplike mandibles with only two teeth, an unconstricted gaster, and a protruding sting. But they also had antlike characteristics such as a reduced, wingless thorax and the petiole segment, or waist, pinched in at front and back, that gives ants their characteristic shape. To the myrmecologist's eye, they looked like ants that owed a great deal to their wasp ancestors. A mysterious, ethereal photograph of this ant in its amber casket made the cover of *Science* and was later used on the book jacket for Wilson's synthesis of the social insects, *The Insect Societies.* Wilson, Brown, and Carpenter were able to define *Sphecomyrma freyi* as a new extinct species as well as part of a whole new ant subfamily (Sphecomyrminae). "Sphecomyrma" means "wasp ant" and "freyi" honors Mr. and Mrs. Edmund Frey, the New Jersey couple who found the fossil and donated it to science. At first Wilson and his colleagues estimated the age of the find to be 100 million years, but this has since been adjusted to 90 million years. This is midway in the Cretaceous period, so these ants or their sisters must occasionally have stung dinosaurs.

Sphecomyrma freyi, according to Wilson, is the "nearly perfect link between some of the modern ants and the nonsocial aculeate wasps," the primitive stinging wasps from which all ants are descended. The antennae were part ant and part wasp, yet these ants had the antibiotic-secreting metapleural gland that is a diagnostic feature found in almost all modern ants. Wilson later reported on additional specimens of *Sphecomyrma* of about the same age from amber in Alberta, Canada. Also, the Russian entomologist G. M. Dlussky has found Upper Cretaceous ants, which Wilson has sorted into two genera, *Sphecomyrma* and *Cretomyrma,* within the subfamily Sphecomyrminae.

BEFORE THE *SPHECOMYRMA* breakthrough, entomologists had looked at ants and other insects in amber for more than a century, but these were mainly from more recent Oligocene and Miocene deposits that shed little light on the ancestry of ants. In the early twentieth century, William M. Wheeler developed a keen interest in fossil ants, especially from Baltic amber. But the specimens he and others looked at and wrote about were members of living subfamilies, and at the generic level they turned out to be the same or similar to the ants alive today.

In general, the evolution of insects is only dimly understood compared to other kinds of life on Earth. Besides the paucity of insect fossils, the reason is partly that the numbers and diversity of species are so much greater than other groups of organisms, and there are many more insect species yet to be discovered. If scientists cannot establish, even to the nearest order of magnitude, the number of insect species, how can we make sense of their evolution?

Insects comprise a class in the phylum Arthropoda, a group that also includes mites, spiders, scorpions, centipedes, and many other organisms. Some 500 million years ago, primitive arthropods such as trilobites and early crustaceans flourished in the sea. But it wasn't until about 390 million years ago, during the Devonian period, that the first insects began appearing in the fossil record.

In the Triassic period, 200 million years ago, roughly at the time the dinosaurs began, one of the younger insect orders, Hymenoptera, which includes bees, wasps, and ants, began appearing. This group would soon rewrite the rules of social living, extending their influence to every corner of the world and becoming much more dominant than all the others.

As the first ants started evolving from nonsocial wasps before about 100 million years ago, wings were a hindrance for these insects of the soil, and most ants lost them. Today, ant workers cannot fly, yet flight remains a crucial aspect of the reproduction of most ant colonies. In most ant species, virgin queens and males have wings, though the females lose theirs before they found a colony. Flight helps ensure gene mixing. In many ant species, newly inseminated queens are able to spread out and populate new territories by flying some distance to found their colonies.

All ants—modern and fossil—are considered to be social insects. The evolution of social behavior began about the time ants appeared at least 100 million years ago. True sociality is found in few insect orders. From looking at living groups, it appears to have evolved only twelve times within the Hymenoptera and once in the line that gave rise to the termites. It has also been found recently in certain beetles and aphids, but it is still uncommon.

The current scenario for ant evolution, reported by Brazilian entomologist C. R. F. Brandão, starts with one Southern Hemisphere ant, believed to be more than 100 million years old. Then, according to Wilson, the next "snapshot" reveals a few species of the primitive Sphecomyrmine ants ranging widely across the supercontinent Laurasia, the protocontinent of the Northern Hemisphere which later broke up into North America, Europe, and Asia. At this time ants were primitive and scarce compared to other insects, according to the numbers found in the Cretaceous amber. But ants already had the antibiotic metapleural gland, which adapted them for life in the soil and, Wilson believes, for true sociality. As these ants began to work the soil, they acquired antlike mandibles and antennae. By the beginning of the Tertiary period, about 65 million years ago, ants began to multiply in number and species, spreading out to every corner of the Earth. There is a primitive ant from the early Eocene, the Fushan deposits of Manchuria, from about 50 million years ago, that reveals traits of both *Sphecomyrma* and the living Ponerinae, the ponerine ants. By the mid-Eocene epoch, about 45 million years ago, according to the amber from Arkansas and Sakhalin (a Russian island off Siberia), many living ant families —including army ants, ponerine ants, and the more advanced myrmicine and formicine ants—were colonizing the Earth. This may have been when the legionary behavior of the army ants evolved. Meanwhile, many other ants evolved distinctive and specialized mandibles and additional poison and other glands.

By the time of the Baltic amber in the Oligocene deposits of 38 million to 23 million years ago, and the Dominican amber in the Miocene deposits 23 million to 7 million years ago, ants had become the most abundant insects, showing that they had begun to dominate the Earth. More than half (56 percent) of the genera from the Baltic amber of the early Oligocene period still survives today. From the early Miocene Dominican amber of about 20 million years ago, 92 percent of the genera survive, and some of the extinct forms are closely related to living species. The surviving Dominican amber species include a miniature army ant, *Neivamyrmex,* the bullet ant, *Paraponera clavata,* and many others found at La Selva. Oddly, army ants no longer live in the Dominican Republic today. In fact, Wilson has found that the current Central American and South American ant fauna is much more similar to the fauna of the area of the Dominican Republic in the Miocene epoch than it is to the current fauna on the island. During Miocene times, the island of Hispaniola, which includes Haiti and the Dominican Republic, may have been a much larger island mass and perhaps even joined to the mainland of the Americas. In any case, the Dominican amber provides the best clues about the evolution of the ant fauna of Costa Rica and the Neotropics.

There are no amber deposits known anywhere in Central America—nothing but Wilson's instant Tanglefoot "fossils"—but Wilson and Brown still dream of finding primitive rarities here. Witness their keen interest in the relatively primitive ponerine or hunting ants, as well as the miracle ant and various army ants. Even if there is no amber around, there is always the chance of finding a living fossil—a crucial link to the past, something alive today and living in an obscure niche, something that somehow stopped evolving millions of years ago, something that will help contribute another puzzle piece to explain the fantastic diversity of ants. If *Sphecomyrma* is a crucial link, a sort of Rosetta stone for ants, then a living primitive ant could be the ant equivalent of the spiny three-eyed tuatara, a large lizardlike reptile and relict of the dinosaur age that lives still on a few small New Zealand islands, or the coelacanth, the last living representative of that group of bony Devonian to Cretaceous fishes, still found at depths in the Indian Ocean. And only living primitive ants can provide insights into early social behavior. Social behavior cannot be deduced from amber unless a whole colony were found preserved in amber—an unlikely event.

Both Wilson and Brown place their primitive ant finds near the top of the list of their myrmecological experiences. When Wilson and Brown start talking about these searches, several countries are mentioned again and again: Madagascar, New Guinea, China, but above all Australia. That most isolated of continents and home of persisting archaic oddities such as the egg-laying mammal, the platypus, and numerous ancient *Eucalyptus* tree species is also home to rare, living primitive ants. Both Brown and Wilson during their *wanderjahren* in the early 1950s visited Australia, keen to find some of the things that Wheeler had written about a few decades earlier as well as things he had missed. And other senior myrmecologists, such as Bert Hölldobler, have made pilgrimages there. The most primitive of Australia's ants, among the most primitive of all living ants, *Nothomyrmecia macrops,* is a large yellow ant that Brown elevated to the status of a solid evolutionary ant prize shortly before he departed for Australia in late 1951. At the time, *Nothomyrmecia* was known from only two workers collected as specimens in alcohol from somewhere near the western end of the Great Australian Bight in 1931 and pronounced a new species in 1934. Brown set out to find living colonies of the ant, following the route of the 1931 expedition, but with no success. Wilson next tried in 1955, searching all the main habitats night and day for a week. No luck. By this time Australian entomologists began to mount their own searches. Neither Brown nor Wilson ever found it, and by the mid-1970s it was feared extinct. In October 1977, as Brown prepared to make an all-out attempt to return to find it, Wilson's former student Robert W. Taylor, an Australian who went to Harvard to get his Ph.D. and then became chief curator of the Australian National Insect Collection, decided the Aussies had better get serious. Taylor led a small convoy of camper trucks on a cross-country expedition west from the capital, Canberra, headed for the Mount Ragged–Esperance area along the southern coast of the state of Western Australia where the only specimens had been found and where so many entomologists had been searching. Along the way, just 350 miles past Adelaide on the long, lonely Eyre Highway, their truck had problems and they decided to camp for the night outside Poochera, a small town in the forbidding Mallee scrub of South Australia. They were still a thousand miles from their destination in Western Australia. It was a chilly spring night with the temperature dropping almost to the fifty-degree mark, a bit too cold for ants much less entomologists to be out wandering. Yet Bob Taylor was too restless to huddle for warmth in the camper truck. Grabbing a flashlight, he broke camp. As Hölldobler and Wilson report in their 1994 book *Journey to the Ants,* a few minutes and just twenty paces away he found something large, yellowish, and antlike out for a stroll along a tree trunk, hunting in the clear, cold night. "The bloody bastard's here," he reportedly yelled as he raced back to the camp. "I've got the *Notho*-bloody-*myrmecia!*"

And so, fittingly, the trophy went to an Australian. The finding, heralded as a "living fossil ant rediscovered" made the pages of *Science.* Since then, Taylor, Hölldobler, and others have studied the behavior of the colonies in the wild, journeying to the ant mecca of Poochera, South Australia, whose small hotel regularly hosts visiting myrmecologists from around the world. But they don't spend much time sleeping; instead they go out hunting for these ants whose niche appears to be their ability to hunt on cold nights when most other ants are inactive. Among other things, researchers have found that *Nothomyrmecia* queens resemble the workers, and all the workers do the same tasks. There are no specialized castes. Each colony has less than a hundred workers. *Nothomyrmecia* huntresses routinely sting their prey, even when the prey is small and could easily be killed in the ant's powerful mandibles. The stinging may help preserve surplus prey, which is stored inside the nest to insulate the colony from hard times. This is a common strategy with ponerine ants. The workers capture flies, bugs, and other insects for their larvae and mainly eat nectar themselves—using two kinds of food like their wasp ancestors. Despite its primitive evolution and its simple social organization, *Nothomyrmecia* lavishes care on its young, just as all other ants do, and stays with them after they reach adulthood. This nurturing social nature is part of the fundamental character of ants. Yet to watch *Nothomyrmecia* in action in the Australian outback is to glimpse what the early social Mesozoic ants must have been like some 100 million years ago or more.

AFTER LUNCH, WILSON and Brown push ever deeper into the forest. They invite an enthusiastic graduate student to join them. David Olson from the University of California, Davis, has spent some months at La Selva collecting ants and other insects. He has wandered through odd corners of this protected forest and knows it well. He is looking at the diversity of ants as an indicator of how diverse the forest may be in general. Hearing of the search for the miracle ant and other primitive ants, and wanting to avoid the area cleared out by the army ant swarm raiders, he suggests they investigate a wild grove of cacao trees—perhaps the remnants of an old farm. Ant colonies abound in the rotting wood and leaf litter there, as well as in the cacao pods.

Three monks now walk through the forest with heads bowed—two senior practitioners and an aspirant. Both Wilson and Brown accept Olson totally and are more eager to learn about his local knowledge than to pass on their own lifetimes of worldwide experience, though Wilson and Brown manage to do both in an offhand, entertaining fashion.

They talk about the growing number of young ant fanatics just entering the field. "It's very encouraging," says Wilson as they walk along. "We'll have people to tell stories to when we're old."

"The thing is to get 'em in time to show 'em a few of the mistakes you can make," says Brown, serious for a moment. "No one ever showed me anything," says Brown. "I wish they had."

"Well, there are just two kinds of people," says Wilson, "the dabblers and the ones that really want to work on ants."

Olson walks along, all ears, but not missing anything on the ground that moves, either.

As they enter cacao country, brown pods hang from odd trees or litter the ground. It is a mixed forest, adorned with strangler figs and lianas, and the small grove of cultivated cacao trees is all but completely overgrown. There are some of the big climax rain forest trees, with classic buttressed trunks, looking like wooden stand-ups for a theater production. There are also, in a light opening, small palms and banana trees with leaves four to six feet long—the daytime homes of tentmaking bats. The bats sleep all day, hanging upside down underneath the leaves. Where these little black bats have roosted, the midribs of the leaves are broken and the leaves are bent in half—like an A-frame tent. Green broken leaves indicate current nests; yellow leaves, recently abandoned nests; and brown leaves, old nests. Stopping to poke around, Wilson breaks up a rotting cacao pod to find three different ant colonies of common rain forest species. They set up the table and portable chairs, and put the tray and the microscope in place. The three fan out in different directions and begin bringing in cacao pods, rotting logs, and leaf litter for closer examination. On the first haul they all converge around the table.

"Well, here we are in the green hell of the Central American jungle," says Wilson, sitting down to start knocking ants into the white tray. "Brave zoologists willing to face army ants, jaguars, and who knows what else. I think the most dangerous thing we've encountered down here is the hot salsa."

"The showers are a little too hot maybe," says Brown. "Almost burned myself this morning."

Olson laughs politely. After nearly stepping on a poisonous bushmaster snake some weeks ago, he has been more careful when leaving the main path and has taken to probing the leaf litter before him with a butterfly hoop net, which, when hoisted high, doubles as a mosquito chaser. Yet for the two veteran antmen who have collected in desperately adverse conditions and would continue to do so for the love of ants if they had to, La Selva resembles a country club. But that is not the main reason they choose to come here. It is because unexpected, bizarre things continually turn up at La Selva. Some of that is simply because tropical lowland rain forests are so diverse, but it may also be because Costa Rica, located near the isthmus of Panama, is near a sort of historical crossroads for many land species. For tens of millions of years of geological history, through most or all of the Tertiary period, North America and Central America were separated from South America. Then, between 3 million and 8 million years ago, with the rise of the isthmus of Panama, species from North America, some of which had walked or been carried across on the land bridge from Asia, began invading South America. And a number of South American species invaded North America. Costa Rica, which had once been at the "end of the road," was now the crossroads and battleground for many of these insect and other animal species.

"These cacao pods are wonderfully easy to split open," says Wilson. "You can hold several whole colonies—each one tiny and primitive—in the palm of one hand."

"Until they start running up your arms carrying their larvae!" says Brown.

While Wilson continues splitting pods, Brown turns to a particularly juicy piece of wood. Meantime, Olson shows them a spider he has found that mimics ants. Two of the eight legs are lifted up to look like antennae. Olson is particularly keen on insect mimics, and Wilson tells him about the ant that acts like a termite that he and Hölldobler found at La Selva. Olson is also collecting a few spiders for his Ph.D. advisor back in California, and he starts telling a spider story. Wilson and Brown are about as excited about spiders as the average housewife. Give them primitive ants any day.

"All right, gentlemen, we have something here," says Wilson. "Now we're talking. This is the way to hunt rare ants. We are looking at—Bill, you'd better be ready to grab this quick—because we are talking *Prionopelta*, the primitive wallpaper ants. And there goes the queen! You want this colony, Bill?"

"Sure," says Brown.

"That's a damn fine colony—if you can get the queen," says Wilson.

"I can't see her," says Brown.

"Over there," says Olson, the youngest pair of eyes.

"See the larvae," says Wilson. "Oh, it's a big colony in here. I see winged forms, too. See the winged queen, Bill?"

"Oh, my," says Brown, too excited to come up with a quip. "A virgin queen."

Her delicate, diaphanous wings are tucked back along her torso. As a virgin, she awaits the coming mating rituals. On that night she will take her turn, flying high, hoping to intercept and mate with potent big-eyed males and found her own colony. But her plans are being changed. She will now be making an unscheduled stop at Cornell University in Ithaca, New York.

"Now that's a proper find in a little piece of log like this," says Wilson.

"These are the ones that Bert Hölldobler and I have studied extensively, the ones that wallpaper their nest. We discovered that they take fragments of their cocoons and plaster them against the walls of their nest to dry it out. It's an entirely new phenomenon for ants. It helps them regulate the temperature and humidity in their nests."

"I haven't read that paper yet," says Brown. "They're little specialized ponerine ants from the amblyoponine tribe, right?"

"Yes, they're primitive ants, Bill, very primitive," says Wilson. "They're found in the Dominican amber, too. And here at La Selva, they specialize on —are you ready for this?—campodeid diplurans. These are little silverfish-like things that most entomologists aren't familiar with, but they're abundant. They're wingless arthropods. Oh, here's one walking around. It's hard to see in this light."

"*Prionopelta* goes for these?" asks Brown, surprised that a primitive ant with rather large colonies could subsist on something so small. "I had them taking geophilids." Geophilidae is a family of centipedes.

"Oh, you've run into wallpaper ants, too?" says Wilson. "Yes, they also take geophilids and a few other things like that, but the things they routinely take and feed to their larvae are these campodeid diplurans. I collected them with my aspirator and laid them out as part of the cafeteria experiments I did here last year with Bert. We got all these arthropods and then just watched what they snacked on."

Olson, who hasn't missed anything, asks, "How primitive are they?"

"They belong to the most primitive living subfamily of ants and one of the most primitive tribes," says Brown.

"Actually, though, this particular species turns out to be not as primitive as we once thought," adds Wilson, a little disappointment in his voice. "Most of the ponerine ants are unassuming hunters living in small colonies, but these actually live in rather large colonies. Anatomically, they seem to represent more of a link between the primitive and some of the more advanced ants."

Wilson and Olson help Brown gather up the workers and some of the nest material and put them in a container. When they are all stashed away safely, Wilson says, "If we can turn up things like that, the miracle ant cannot be far away."

A few more hours of searching, however, produce no miracle ants, although there are a number of other worthwhile finds. Wilson suggests that they take the long looping trail back to camp for dinner and check on the Tanglefoot traps along the way. They gather up their finds, stuffing pockets and knapsacks full of new colonies and specimens for future study.

The late-afternoon stroll through the forest produces more talk of evolution and amber, about the many surprises that a Central American rain forest holds for modern-day Darwinists. "You'll have to come up and see my bala fossils sometime," says Wilson. "They're in huge pieces of Dominican amber. I paid one hundred dollars each for them because they're so large."

At the first Tanglefoot trap, Wilson finds some robber, or asilid, flies. Lower down on the same tree he sees an aztec ant caught inside another trap, perhaps trying to extend the range of a nearby ant–cecropia tree colony.

"This stuff is really working," Wilson says. Then he notices more aztecs putting dirt and leaves on the Tanglefoot, all around their fallen nestmate. Wilson had observed a similar thing in his preliminary experiments with Tanglefoot back in his Harvard lab. When he put Tanglefoot out for his lab ants, after the first ant got caught, the other workers would bring worker corpses from the refuse heap to pile onto the Tanglefoot. It effectively kept the rest of the colony from wandering into the sticky stuff. Maybe some of the ants in amber are really corpses that the ants deposited there millions of years ago.

The next tree, fifty feet down the path, yields gold.

"It's *Acanthoponera,* Bill!" Wilson yells.

Brown and Olson are poking through the leaf litter and don't appear to hear at first. Wilson leans down with his hand magnifier to study what turns out to be a winged ant trapped in Tanglefoot. He assumes an intense expression, content for the moment to be alone with the treasure. He is at once back in his home lapidary shop in suburban Massachusetts, engaged in what he calls "the most delightful occupation in the world," polishing amber fossils, putting them under the microscope, and turning them over and over, searching for the clearest view that will give him the definitive peek into the Earth's distant past. His reverie is broken by the swishing leaves signifying the approach of his two companions.

"It's an *Acanthoponera* queen," says Wilson. "I've seen only two *Acanthoponera* the whole time I've been visiting Costa Rica."

"*Acanthoponera* is another primitive ponerine," Brown explains to Olson. "It's from the tribe Ectatommini."

"This is another of Bill's specialties that he has sorted out," says Wilson. "Look at this! A queen just flew in and got trapped, and now she's sinking in. If that were tree gum, she'd be almost embedded."

Brown takes a closer look with the jeweler's loupe: "It's *Acanthoponera minor.* Can you get that stuff off?"

"With great difficulty, if at all. I haven't worried about that part of it. You want the queen, huh?"

"That's a nice specimen," says Brown. "We could throw it in alcohol now and then find out the solvent later."

"I'm sure we can find something," says Wilson, "but that's not really the point. I'm not interested in collecting this way. I'm interested in finding out what genera and species of ants are caught and in what proportions."

Brown drops the matter and they walk on, Wilson stopping to examine more traps. But the question about the solvent for Tanglefoot soon arises again with greater urgency as Wilson absentmindedly brushes against one of his own traps. It happens only seconds after he jokingly warns Olson, "be careful, or I'll have to record you if you get tangled up in this."

After a few sheepish moments of silence, Wilson finally admits his difficulty. "I'm making a true mess of myself with this Tanglefoot. How will I get this off?" Instinctively, he tries with his one free hand—and then backs into even more of the stuff. It soon spreads to his trousers and knapsack. Wilson notes out loud: "Must warn about the future use of Tanglefoot."

One of the last traps offers even more surprises. A four-inch long lizard is caught in the Tanglefoot, alternately writhing and holding still. Only the lizard's feet are stuck—at least for the moment. Wilson tries to set it free: "There is a case of an eight-inch lizard caught in the Dominican amber," he says. He finally succeeds in releasing it.

He walks on, joining the others now as they press on toward the camp. Could the ants or other primitive insects, preserved in amber, ever be brought back to life? It's a question Wilson is sometimes asked. In *Jurassic Park,* the best-selling thriller made into a popular Hollywood movie, Michael Crichton invents the story of scientists who are able to extract dinosaur DNA from blood samples taken from the stomachs of insects caught in amber. Using this DNA, the scientists reconstruct dinosaurs.

Around the time of the film's release in 1993, there were many sensational "science" stories that intimated this might be scientifically possible. In fact, since the film was made, two teams of American scientists have extracted DNA from two 30-million-year-old insects trapped in amber. They recovered the gene fragment of a termite and a stingless bee species, both long extinct. The research is important because these gene fragments can now be compared to the insects' modern relatives to determine how closely they are related.

But re-creating a whole insect, not to mention a dinosaur, is far-fetched. There wasn't even enough DNA to re-create a single gene. On the near-scientific horizon, there is some promise of being able to re-create simple bacteria or single-cell organisms, but anything larger or more complex would be very difficult.

Moreover, the idea of extracting blood from an amber insect's stomach reveals a lack of understanding about the nature of an insect fossil in amber.

For the most part, the amber captures just the impression of the insect exoskeleton. There is little or no interior. As Wilson grinds and polishes, he simply cuts a new face on the gem, usually to see the front of the ant's head or to glimpse the teeth on the mandibles or other distinguishing features. In some cases, the amber must be ground down to within a few thousandths of an inch to see clearly. But if you get to the ant itself, says Wilson, "you discover that, like the smile of the Cheshire cat, it disappears. There is no ant there." After the ant became trapped, the amber solidified and the ant decomposed, leaving nothing but a carbon film and sometimes a few other more complex molecules. The ant is essentially a cavity. In a few cases, bits of dehydrated tissue have been found preserved—fragments of muscle as well as cuticle. Yet even if the blood of dinosaurs could be found and DNA extracted, rebuilding a dinosaur would be a task similar to creating a human from a sliver of a fingernail—a job best left to a Hollywood special effects department.

Even if primitive ants could be brought back to life, who would want to do it? Ants, according to Wilson, are among the strongest candidates for a group of organisms that will outlast humans. (The overwhelming odds are that humans will eventually become extinct, as have 99 percent of all organisms on Earth.) The study of ant fossils and evolution tells us many things, but it confirms one thing with absolute clarity: Ants are not only supremely successful today, but they have been successful for a long, long time. If you define success as predominance in numbers and territory, then ants may be the most successful terrestrial creatures that have ever lived. Despite concerted effort, campaigns to eradicate such ants as the imported fire ant, *Solenopsis invicta*, have met with individual successes from limited areas in the American South but overall failure because the ants have continued to spread, evolving new strategies such as multiqueen colonies. Ants are also resistant to hard radiation. In a French forest, ant colonies were exposed to intense cesium-based irradiation over eleven months but suffered no evident decline or change in behavior —even after some of the surrounding plants started dying or losing their leaves. It is often said that cockroaches will inherit the Earth, but instead the inheritors may be ants.

As the three men leave the forest, the leafcutter ants are out foraging. The long trail of leaf carriers parallels their path for the last hundred yards as the ants move to and from their nest at the edge of the forest near the overgrown plantation along the Río Puerto Viejo. After all the small primitive hunting ants with one or two castes, it's like coming upon a Mayan civilization in the jungle—complete with pyramids and gardens and successful cities of up to 7 million inhabitants going about their complex business. It's mesmerizing to watch—an ancient civilization at work. Then Olson, leading the way, stops. As he points across the path from the leafcutters, the two senior myrmecologists almost trip over each other.

"It's *Apterostigma,* I think," says Olson.

"It is indeed," says Wilson. "They're primitive fungus ants. They don't even cut leaves but mainly forage for seeds, flower stamens, insect droppings and other organic detritus to fertilize their gardens. That eliminates the need for a few castes: the ones that in other species cut and carry the leaves and dice them into tiny pieces."

In the early development of all leafcutter ant colonies, no matter how advanced, a queen with no workers cultivates the fungus gardens herself, mainly using her feces. This primitive trait can be seen as a permanent feature of these tiny fungus ants. *Apterostigma* are not primitive ants among ants generally, but they are one of the more primitive species of the highly advanced tribe of the fungus ants. Observed beside the vast, sprawling, highly evolved leafcutters, the primitive fungus growers are tiny individuals as well as small in colony size. Their garden is pocket-sized. It provides a dramatic example of the evolution of ants and the wide gulf between the primitive and the advanced. All ants are social, but some are a good deal more social than others.

Chapter Five

Coming of Age

Their . . . exquisitely cooperating workers . . . conceal . . . the nuptials of their strange, fertile castes, and the rearing of their young, in the inaccessible penetralia of the soil.

—*William Morton Wheeler,*

ANTS, *1910*

The uniformity of Earth's life, more astonishing than its diversity, is accountable by the high probability that we derived originally from some single cell, fertilized in a bolt of lightning as the earth cooled. It is from the progeny of this parent cell that we all take our looks; we still share genes around, and the resemblance of the enzymes of grasses to those of whales is in fact a family resemblance.

—*Lewis Thomas,*

THE LIVES OF A CELL, *1974*

The ant colony is essentially a factory within a fortress, a splendid arrangement of soldiers, builders, nurses, and other specialists united in single-minded dedication to the survival and reproduction of the queen.

—*Edward O. Wilson,*

FROM AN INTERVIEW, *1990*

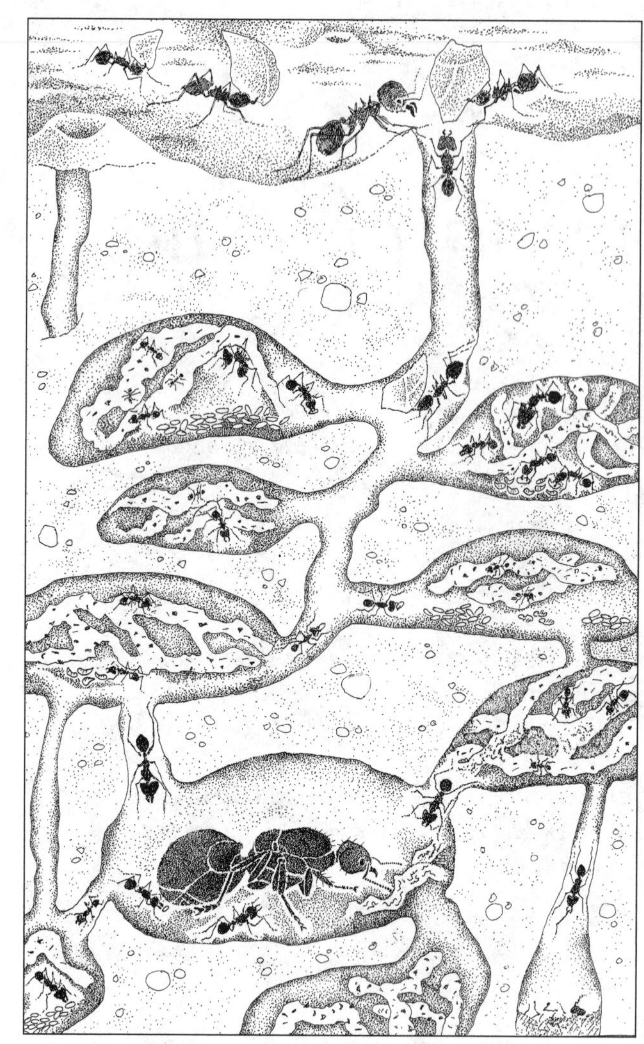

The Queendom of Leafcutters, Atta cephalotes. *At top, the scout and other foragers bring leaves back to the nest, passing a huge soldier on guard. The foragers take the leaves to a garden chamber where the smaller shredder ants cut them into tiny pieces and pass them on to the even smaller pellet-makers, who mold them into moist pellets and slip them into the mushroom gardens. Transplanters move the fungus from chamber to chamber while the harvesters, or minimas, weed and harvest the ripe fungus for the colony. The queen —about the size of a baby mouse—lies at the deep heart of the nest.*

EVEN the low rumbles from ten to twenty feet overhead—humans tramping across the roof of the colony—cannot disturb the intensive activity. Deep inside the leafcutter nest, the mother queen, reclining in her chambers, pushes out another egg from the rear of her swollen gaster. For an ant she is massive, and the golden gaster of this baby mouse–sized ant is stuffed with thousands of eggs.

The eggs emerge one at a time, a few minutes apart, though some arrive in strands of three or four. Each egg—pure white and elliptical like a chicken egg—is one-sixty-fourth of an inch long, about the size of a grain of sand. The queen's attendants—part of the worker caste that mainly do the pellet-molding—surround her, wait on her, groom her, and protect her, removing her eggs one by one the instant they are laid or a second or two later as they drop to the chamber floor. The attendants lick the eggs to keep them clean and to help confer on them the colony odor. Then the workers transport them to adjoining rooms to slip them into piles of growing fungi that will provide food as the eggs hatch into larvae. As well as keeping the chambers clean, the workers monitor the temperature and humidity. When one room's conditions are less than ideal, the workers move the eggs, fungus and all, to other chambers. For the next approximately seven weeks, as the eggs turn first into larvae and then pupae before hatching into adults, the workers will dote on these developing ants.

Over the previous two and a half years, the mother queen has steadily pushed out eggs, more than a million a year. All, to date, have been workers—various harvesters, transplanters, pellet-makers, shredders, foragers, and others, up to the large soldiers. All are genetically females but unable to reproduce. It takes about forty to sixty days for a colony to rear a new brood of workers. In her lifetime of up to fifteen or twenty years, the leafcutter queen will produce 20 million or more eggs that will hatch and grow mostly into worker ants.

But the next eggs she lays will be special—her first royal brood, the first time she has produced males and virgin queens. These eggs will look the same as the eggs that hatch into workers, but the workers charged with caring for them will treat them differently. Soon the message will be on this brood for all the colony to taste: Take special care, for this is our ticket to immortality. These will be ants born with wings to fly away, mate, and found new colonies. The workers who will care for this reproductive brood are already buzzing, getting ready.

In recent days the leafcutter scout, too, has been busy. Her efforts to locate new leaf sources and to direct the dismantling of trees, bushes, and bean patches has proceeded at an almost frantic pace. After a few nights of rain a week ago, the new growth in the rain forest is luxuriant, especially near the tree falls and light openings, and the ground is dry, ideal for collecting. The scout could not have found things more suitable for the coming royal brood. Meantime, inside the colony, other workers have removed or turned over the substrate from many old gardens, taking the scout and her associates' offerings of fresh leaves to start the next crop of fungi. Still other workers in the colony have been working night and day to excavate new rooms and expand the size of old ones.

After almost three years, the number of garden chambers now stands at 212.

There are also a few large waste chambers, or garbage dumps, up to three feet by four feet, excavated to hold dead ants, and exhausted substrate from the gardens and other waste materials. The dumps, which draw various flies as well as many other insects and microorganisms, must be kept separate to avoid contamination of the brood and the fungus gardens. A common sight has been the worker ants gathering outside the dumps, spitting out pellets of waste material, while the flies below buzz around, laying new eggs on choice morsels.

Working simultaneously in many different areas around the colony, the workers have been able to achieve substantial progress in their preparations for the new queens and males. This is how the pyramids were built in Central America and Egypt. In the leafcutter colony, it is how a superorganism moves mountains of soil and vegetation and raises its gardens as it prepares for the most important phase of its life and the ultimate reason for living: reproduction. As do humans, ants build their homes using bold plans and designs with permanence in mind.

In ants as in other organisms, the period of the coming of age can be an awkward time. The colony first has to raise healthy virgin queens and males. It must give them special attention compared to all other broods—more food and nurturing. Getting them out into the world and giving them a fair chance is the colony's challenge. Many obstacles can arise: heavy rains, even floods, which prevent the ants from leaving the nest, or drought, which reduces the supply of fresh green leaves. There can be attacks from rooting anteaters or nearby colonies of swarm and column raiders just as the reproductives are leaving the nest. There are the ever persistent phorid flies, the tiny flies that bother many species of ants, laying eggs that parasitize and kill the ant hosts. But most of these hazards present themselves only outside the nest. Life at home in the colony is mostly safe. And with ants, obstacles to colony success are generally challenges quickly met and conquered.

In another corner of the La Selva forest, only one hundred yards away, the swarm raider queen reclines in her protective bivouac. Like the leafcutter queen, she has also continually produced new worker broods—in this case for a year. And now her first reproductive eggs are about to be laid. As the colony prepares to rear its queens and young males, things are surprisingly quiet. Day-to-day life for army ants is fundamentally different from that of leafcutters, and so is their reproductive life; for example, the army ant queen's egg laying has been on more complex cycles.

Most army ant species alternate between two cycles—the statary, or stationary, cycle and the nomadic, or migratory, cycle—throughout the entire life span of their colonies. During the stationary cycle, the colony stays close to home—the bivouac—sending out only minor hunting parties on the odd day.

The bivouac sometimes moves slightly but largely remains stationary. With the queen and her attendants is the brood born during the previous stationary phase about a month earlier, having just changed from wriggling larvae to fairly inert pupae. But they are not the focus of attention. The focus is the queen herself, with her gaster swelling larger day by day. Toward the middle of the nineteen- to twenty-two-day stationary cycle the queen will lay her eggs in a single series—thousands of them.

The number of eggs produced by a swarm raider queen is on the order of 2 million per year. The fastest rate of egg production of any ant species in the Neotropical jungle, this works out to about one hundred thousand to three hundred thousand eggs per brood. The swarm raiders' royal brood is much smaller, however—little more than three thousand eggs. Some ant queens with smaller colonies produce as few as one hundred eggs per year. But the record for all ants, and probably for all insects, is 50 million per year in African driver army ants, *Dorylus nigricans.* Their colonies are up to 20 million strong, twenty times as large as their Latin American relations, the swarm raiders.

A few days later when the eggs in the royal brood hatch into hungry larvae, they receive the devoted attention of the workers. Several days after that, the previous brood, now maturing pupae, begins to emerge as callow workers—new, pale-colored adult ants. They are callow as they first emerge, their exoskeletons still soft and unpigmented. With two broods maturing at different times, the workers in the bivouac must divide their attentions.

As the callow adults become full-fledged workers over the next few days and the larvae remain ever hungry, the colony turns to its migratory phase. By day the colony now dispatches large hunting parties of the swarm raiders in search of food. There are hundreds of thousands of new mouths to feed—new workers as well as hungry larvae. As each hunting day ends, after night falls, the colony moves the bivouac. Surrounded by her workers and entourage, the queen walks to the new site. The new workers join in the full activity of the colony, some of them helping to carry the hungry larvae on the nightly migrations. For fourteen to twenty-one days and nights the swarm raiders raise hell through the jungle, as if they own the place. They more or less do.

After that, as the royal larvae metamorphose into pupae and the colony again settles back into another stationary cycle, the queen gets ready to lay even more eggs. At the end of this second stationary phase, the royals—the virgin queens and males—turn into callow and then full adults. It takes a total of thirty-one to thirty-four days for them to develop.

Early naturalists assumed that army ant movements were triggered by a colony's need to find new raiding territories. Army ants can quickly deplete local areas and certainly need to keep moving on. But the trigger for the behavioral changes of the colony in the two alternating cycles is the activity level of the brood. In 1932, on his first trip to study army ants in Panama, Theodore Schneirla worked this out, and on many later trips he filled in the details. The high activity levels of the larvae and the emergence of the callow workers trigger the migratory phase. After the larvae become pupae, their inertness signals the time for the depressed activity levels of the stationary phase. Then the queen, too big to move, will focus on laying her eggs. Because it takes more than two cycles for a brood to develop and leave the nest, two broods overlap each other in time as well as space.

But as Bert Hölldobler and E. O. Wilson have pointed out, Schneirla virtually ignored until late in his research the important role that the need for food plays in army ant migrations. Yes, the *immediate* cause of migration is the behavior of the larvae, but the *ultimate* cause, the adaptive value, is that more food can be had by the colony when it moves regularly. Biological systems, as Hölldobler and Wilson report, often evolve to make use of regular rhythms in the life of an organism, in this case the emergence of callow workers, because then they do not have to rely on a close daily reading of their environment. The superorganism "knows" that when callow workers appear that's the signal to move on. And by following that "rule" they will have enough food on the table.

Both the leafcutter colony and the swarm-raiding army ant colony at La Selva have now reached the momentous time of the coming of age. For months or even several years ant superorganisms invest all their resources in growth, reaching totals of more than a million workers in the leafcutters and almost a million in the swarm raiders. But eventually it is time for these colonies to pursue their ultimate goal, the goal that drives every living organism on Earth: to reproduce and to send one's genes into the next generation.

The stage of initial growth leading to sexual maturity followed by reproduction is found in many organisms. In flowering plants, an annual will spend its first few months growing strong stalks and making leaves—the food factories of the plant; midway or late in the season, reproductive activities begin. Then the plant turns to making flowers, fruit, and seeds, before dying. A tree or other perennial plant operates on a similar agenda, although the process may be spread out over several years or even decades. A human can spend a dozen or more years growing before turning to the reproductive stage of life.

Once the reproductive stage is reached, organisms typically slow down or stop their growth. All or most of their energy is now devoted to reproduction. In some organisms, such as salmon, reproduction is a one-time-only event. In others, such as mammals and ants, reproduction occurs seasonally or often throughout the year and over a period of years.

The leafcutters and the swarm raiders are only two ant colonies of many thousands in the surrounding forest at La Selva. Every day, especially through spring and summer, hundreds of ant colonies of all sizes are rearing males and young queens and preparing to send them out to mate and reproduce. A given species may use different parts of the year, but the best strategy is usually when colonies of the same species send their queens and males out to mate around the same time. This ensures a choice of potential mates so that the genes are mixed. The fittest of both sexes tend to succeed—the strongest fliers, the most ardent and durable lovemakers, the hardiest individuals. But it is not just a case of genetics. The fittest ants for mating will also be those that received the best care and nourishment through the egg, larval, and pupal stages in the nest. Part of what is ultimately being tested is the fitness of the mothering colony. In the queendom of the ants, new queens and their male suitors are not just born but have to be made.

Week One

On two successive mornings the army ant swarm raiders attack an extended arm of the foraging leafcutters. The army ants seem highly charged, aggressive, at times fierce, like adolescent boys with testosterone racing through their bodies. The leafcutters are rugged, capable ants, and their soldiers inflict many army ant casualties. Still, the swarm raiders' sheer hyperaggressiveness and their massive raiding parties give them the edge. Also, both times the army ants have the advantage of a surprise attack. But these are skirmishes, not wars; and neither colony feels much impact. Then as suddenly as they started, the swarm raids stop; now the superorganism is subdued, mostly quiet. One day a feeble hunting party rounds up a few odd arthropods, then all is silent for several more days. Something is happening inside the bivouac at the heart of the colony.

Meantime, inside the leafcutters' royal chamber, the mother queen rests, slowed down by her swollen condition, her baby-mouse size, preparing to lay the most important eggs of her young life. Seconds earlier, inside her gaster, the mother queen released a minute amount of sperm to join with an egg and fertilize it. The first new candidate leafcutter queen is conceived. In fact, to the queen it is just another worker female. The sperm, one of 200 million to 300 million that she has been using to make all the workers, came from the spermatheca, a whitish bladder below the rear of her gut. Ever since the queen mated a few years ago, this sperm has been held as free-living cells, nourished and maintained in the spermatheca and dispensed as needed to make females. It will last fifteen or twenty years, the active egg-laying life of a leafcutter queen.

The mother queen rocks her gaster forward, then back, then forward again. The undulations ease the egg down the fallopian tube and out into the world. A worker is there waiting to catch it in her mandibles.

The egg of the prospective virgin leafcutter queen is indistinguishable from the eggs that will metamorphose into workers or males. The same size egg produces all the castes of leafcutters, from the smallest worker to the soldier and even up to the giant queen, who at full size, laden with eggs, is eight hundred times the weight of her smallest worker.

A few minutes later the mother queen produces a string of eggs that are all males. These are not fertilized with sperm. As they are dropped on the floor of her chamber, they are carted off by workers to various chambers.

Over a period of a week, thousands of leafcutter males and virgin leafcutter queens are laid as eggs. There are more than seven times as many males as females; by one count for leafcutters, there are thirty-eight thousand males to five thousand females. All are repeatedly licked and then carefully wedged into the fungus gardens in a number of rooms quickly turned into "ant nurseries." The workers will now decide the fate of this royal brood; it will be up to them to turn these ordinary eggs into a royal brood.

Week Two

In the leafcutter nest, in one of the "nurseries" of the prospective queens and males, the tiny kohlrabi heads of the fungus now all but obscure the virgin queen's egg, which has become embedded in the grayish mass. The threadlike mycelia form a dense network that resembles a woven wall-hanging covering the mass of queen and male eggs. In places the weave is tight and in others rather loose, with the eggs just peeking out. On two successive days the workers sense the temperature of the room rising too fast and the eggs and the fungus drying out. They move the eggs containing the virgin queens and males, all with the fungus attached, to another room deeper in the nest, closer to the water table, slightly cooler and distinctly damper. It takes sixty trips by the sixteen ants then on duty in the nursery to move all three hundred eggs. At the same time, in a dozen nearby rooms, the eggs are also moved. In other rooms, the temperature stays within an acceptable range.

By midweek, aboveground and a hundred yards away in the army ant colony, six fertilized eggs are laid in the warm center of the bivouac—the first eggs that will become army ant queens. It is a different scene from that in the leafcutters' colony, yet the aim—reproduction—is the same. The eggs cannot be moved from room to room as with the leafcutters, so the army ants change the temperature or humidity of the "room"—the bivouac itself. As the attendant workers lick the eggs, the workers around the walls close ranks to raise the temperature, turning the queen's refuge into an incubator and maternity ward.

A few days later the mother queen army ant lays a second set of eggs, this time unfertilized. These eggs contain males—some three thousand of them—a large number compared to the six queens, but the total reproductive brood, three thousand and six, is small compared to the usual all-worker broods that can number three hundred thousand. In these massive broods, competition for food is fierce. The small size of the royal broods, by comparison, ensures more food and attention to each virgin queen and male—care that is crucial to their development. In addition, they receive special nutrients from the workers. Overfed and well tended, each queen and male will attain a size much larger than the average worker. Their forms are not just larger but unique. With the larval virgin queens, part of their overfeeding may be special nutrients that are thought to be produced in one of the exocrine glands in the head or thorax of the workers.

It was Wheeler, reporting from British Guiana in 1921, who first discovered the virgin queens of the swarm-raiding army ants. He found two callow queens among hundreds of male cocoons in a bivouac. He saw that the males and potential queens of army ants were reared together and that these queens, unlike most other virgin ant queens had no wings, and therefore no nuptial flight.

For the army ant colony, this is the first annual royal or sexual brood. Most ant species produce their royal broods on a regular, often annual, basis, in the spring or summer, when food resources are best. Worker broods are usually produced continuously during the rest of the year. Schneirla found that most swarm raider royal broods start during first half of the dry season, early in the calendar year. Some royal broods, however, occur in rainy and other seasons according to local conditions. A bivouac located in low, wet surface conditions may not contain a royal brood even in the dry season, whereas a bivouac situated on high ground in a dry area of the forest could start a royal brood even during the rainy season.

In all ant species, the caste of the brood—whether an egg turns out to be a soldier, a tiny worker, or a virgin queen—is determined by the food, mainly how much a larva is fed and/or the special nutrients received from the workers. The workers therefore determine the caste, based on what they are able to learn abut the needs of the colony. The queen determines the sex by releasing or withholding stored sperm from a pouch in her gaster to fertilize each egg before she lays it. The vast majority of her offspring will be females—from fertilized eggs—so there is not much to be determined. It is only for royal or sexual broods that she must produce some males. For these, she simply withholds sperm, so the eggs remain unfertilized. In some species, even virgin sister workers can lay male eggs—though they must then compete for success with the queen's male offspring, which may be larger, more robust, and better cared for.

Weeks Three and Four

As the leafcutter eggs mature, the embryos inside shrink a little and rotate. In the egg that holds the larval virgin queen, a ghostly visage slowly emerges, headfirst. Her mouth parts opening and closing, she thrusts her head capsule against the shell. Rasping her spiny mandibles against the jagged edge, she gradually gnaws a big hole through the side. After a few minutes of vigorous effort, the larval queen hatches. The driving force for this determined escape act, according to one ant scientist, may simply be the reflexes of hungry larvae.

The larval leafcutter, as in most ant species, has no eyes or legs. The larvae of all gardening ants show a characteristic plumpness when viewed from the side, with the head curved around and facing backward. Myrmecologists who specialize in studying larvae have found these and a few other distinctive differences when comparing gardening ants to other ant tribes, as well as subtle differences among the various species of gardening ants. As a leafcutter species, the larval virgin queen starts out looking like any larva from her colony. Yet as she is fed more and more by the worker ants, who spend so much time attending her, she begins to acquire a subtle yet special, powerful attractiveness. All the larvae are licked often, but the larval queens get the most attention. As saliva is added to the integument, or outer coat, of the larvae, the fungus, their food supply, grows all around them and over them.

The most important feature of all larvae is the mouth, usually gaping. Some ant larvae are able to wriggle to food, but leafcutter larvae have to be fed even though they are surrounded by the growing fungus and embedded in it in the early larval stages. Sometimes all that can be seen through the fungus is the larva's head, its short hair bristles around the mouth forming a sort of basket in which the workers can drop the fungus during feeding time.

As the larval virgin queen and several other larvae move their mouths, obviously hungry, a devoted worker is already picking the inflated kohlrabi tips of the fungus. She brings three of them to the larval queen, licking her several times before and after feeding. Even then the larva remains unsatisfied. The worker brings more and more. Other workers help, too. The larval queen is growing quickly; some larvae can double their bulk in one day. Within a week she is already many times larger than the egg from which she emerged.

Over in the army ant bivouac, as the sun peeks through the lattice of worker legs and bodies that form the walls, the six army ant virgin queens hatch into larvae first, followed a few days later by the males. Overnight the bivouac incubator that is the army ant colony becomes a nursery and feeding center.

Everywhere, nurse workers are coming and going as in a ward in a metropolitan hospital. The workers crowd around the larvae, licking them and offering them pellets of food, which are placed right before them. The pellets are the rolled-up soft parts of insects—the booty from recent swarm raids—chewed up and blended into "baby food" by the workers. Right from the start the army ants are reared as carnivores compared to the leafcutters, which are fungivores. Both broods, however, are at least sometimes offered alimentary, or food, eggs. The army ant queen lays these eggs right after the males are laid. The food eggs are soft and can be easily squeezed open by the workers and fed as nutritious supplements to the reproductive brood.

Army ant larvae are whitish, long, and slender, with the front end slightly curved. Their comparatively large head has only weakly developed mandibles and other mouth parts. They have conspicuous vestigial legs—conspicuous at least in relation to other ant larvae. Even to the myrmecologist, they're still just microscopic pimples. Their mouths, ever open—as are the mouths of leafcutter larvae—solicit the workers for food. Even though the army ant royal brood is only 1 percent the size of a typical worker brood, it will take more total food to feed it. The reproductive larvae have "unique secretory properties," as army ant specialist Schneirla put it, which ensure their intense attractiveness to the workers, who give them much more attention than to the usual worker broods.

By the end of the third week the army ant colony becomes restless. The larvae have worked it up to a near frenzy. Early on a muggy morning in April, the first big swarm raids in several weeks depart from the bivouac to scour the jungle. It is the fiercest swarm raid this colony has ever staged. They take thousands of surprised ants of the sturdy species *Gnamptogenys* as well as the rugged little fire ants, various elegant dacetines all. They even take a few rare miracle ants, with their pitchfork-shaped mandibles. Several La Selva researchers, busy collecting that morning, happen to see the big raid but miss the miracle ant. Wilson and Brown have returned respectively to Harvard and Cornell, and only David Olson continues actively to look for the miracle ant. The army ants have no trouble finding it.

At day's end, as the swarm raiders return to the bivouac, the colony undertakes its first nightly migration of the migratory phase with its reproductive brood. Everything—mother queen, brood, and food supplies—must be moved. The tender army ant larvae, vulnerable to cooling, are carried in groups packed together, which tends to keep them warm. The mother queen walks, surrounded by her protectors. She is followed by her extensive entourage and, pulling up the rear, the various camp-following insects and other arthropods that hang out with army ants.

Weeks Five and Six

Midway through week five, as the leafcutter larvae mature, pupation becomes a colony effort. Several adult workers surround the larval virgin queen and remove the larval envelope—which is actually the integument of the last larval stage—to reveal her naked, all white pupa. Her legs and wings are free but pressed tightly to the rest of her body. Her mandibles and antennae are pointing down and toward the rear underside of her developing body. The rear pair of legs extends beyond the body.

The eyes of the pupal virgin queen begin to darken, making a stark contrast to her pale white body. Later in the week the edges of the mandibles darken, and then the entire body turns pale brown and finally a dark reddish brown.

Up at ground level, day after day, the army ants continue their raids, moving their bivouac every night as they have for almost two weeks. As the larvae grow larger, about three thousand workers, one for each larva, are needed to carry the brood during the nightly migrations. Each worker grasps her designated larva around the middle, holding it almost as a monkey would its young, belly to belly in a slung-beneath position. This also enables the worker to immobilize the larva by pressing on its most sensitive region, where the main neural ganglion is located. The worker's mouth remains in close contact with the larva to provide food as well as social comfort, stimulation, and communication. Schneirla felt that such constant intimate contact "must aid greatly in forming the social bonds that weld new individuals into their colony." But with the royal brood, it is also the wedge that will soon split the colony in half.

The perfume of the larval virgin queens is beginning to affect the swarm-raider workers, their sisters. Each night that the reproductive brood is carried on to the next bivouac, it is placed farther and farther away from the old mother queen. In the bivouac, more and more workers begin to cluster around the new brood, paying them oral homage and offering special favors. The virgin larval queens are kept along one side of the bivouac, each surrounded by a growing number of attendants, while the queen remains on the other side of the nest with her loyalists. As this polarized situation persists, workers decide either to remain loyal or to switch their allegiance from their mother queen to one or another of these developing queens. It is a potentially explosive development for the colony—a ticking time bomb.

Sometimes the leafcutters below in their nest feel the swarm raiders thundering above—the awesome sound of pounding feet from so many Huns and Tatars. Several times the raids pass directly overhead. The leafcutters then seal their entrance holes or station soldiers to stand guard just inside the entrance. Then one night the army ants move their bivouac almost on top of the leafcutter colony. They are only a few feet away on the riverbank. Army ants "in bivouac" are not usually dangerous, but the leafcutters respond by dispatching their soldiers and putting the colony on full alert, ready to defend and/or evacuate if necessary.

The next morning, as the army ants send out the dawn raiding party, the leafcutters remain below. A couple of hours later when the army ants are mostly gone, the scout slips out past the army ant bivouac, laying a pheromone trail to the day's leaf harvesting. A foraging column of leafcutters follows soon after, many of whom refresh the trail regularly as they go along. Their adaptability to the situation—staying down and then avoiding the army ants—saves many lives and considerable energy.

Through week five the army ant larvae grow larger and heavier, and require more effort to move every night. The queen larvae are especially heavy. Both the larval queens and the males are about an inch long, but the queens are more robust—almost double the girth of the males. The bulging larvae require even more food. The rolled-up food pellets arrive fresh from the daily raiding parties, but more and more are needed to satisfy each larva. Also, because of the traveling, the larvae must be fed extra pellets both before and after the journey.

As the migratory cycle nears its end, the workers carrying the queen larvae trail behind all the others, transferring them only after most of the others have left. Each larva is carried by a submajor, or smaller, worker running in tandem in the midst of a small crowd of faithful workers. During the transport there is no conflict among the various groups of queen larvae and worker attendants, but one night when they are back in the bivouac, combat breaks out. Workers of two groups are scrambling for a favorite position in the bivouac, and the jostling soon turns to shoving, then biting, and finally killing. When the fight is over, six workers lie dead. As workers adjust their positions each night in the bivouac, such combat becomes common. In the nights that follow, more workers die as the colony divides into even more factions, and one night rival workers manage to sting a larval queen, fatally injuring her.

Early one morning, after the colony moves for the eleventh day in a row, one of the larval queens matures and begins to spin her own cocoon. At first her weaving presents no more than a tissue-thin transparent curtain. Through it, the larval queen can be observed as she spins. Gradually, the cocoon turns darker and her icy image disappears. The cocoon becomes a shroud, a mummy case.

Among the swarm raiders, males also spin cocoons, and females spin them whether part of royal broods or not. In the *Neivamyrmex* miniature army ants, only the royal broods spin cocoons, the worker broods remain naked. None of the leafcutter ants spin them. The ant cocoon is inherited from the wasplike ancestor of ants; therefore, the absence of a cocoon is an evolved characteristic. The significance or advantage of having a cocoon, if any, is unknown.

At the end of week five, all the naked queens and males are spinning their cocoons—all arranged in separate groups on the opposite side of the bivouac from their mother queen. Their estrangement, crucial to reproduction, is almost complete. But the bivouac has become a tense, difficult place; the colony can barely be called a "superorganism." Another virgin queen and several male pupae placed too near the mother queen's entourage are taken and eaten by several workers loyal to the queen. All that's left is the spent pupal casing ripped open like a birthday present. This kind of cannibalism occurs at such a low frequency that it never jeopardizes the survival of most of the reproductive brood, including the most attractive queens and males.

The colony moves one last time in the migratory cycle and then settles down to begin the next stationary cycle. With the hungry larvae now bundled up in their cocoons—their mouths effectively gagged—the workers who surround the brood can rest a little in week six. There is plenty of food for the mother queen. Her loyalists bring more and more fresh pellets. She needs them. She is about to lay several hundred thousand eggs for the next brood of workers. During the previous nomadic cycle, many workers died of natural or other causes—some in skirmishes protecting larvae—and even more have switched their allegiance to the virgins. Many more workers will be needed to fill the ranks and carry on the mother colony.

Schneirla first observed the division of a swarm raider army ant colony in Panama in 1946. In one experiment he heated the colony with reflecting trays to render it listless, then opened it up, peeling away the lattice of worker legs to reveal the bipolar positioning of the mother queen and her estranged virgin daughter queens. A day after the nomadic phase, with the sexual larvae newly hatched, he removed the mother queen to see what would happen.

Schneirla returned her to the colony after three hours. The queen was accepted and indeed did not even seem to be missed. Ten days later, after fierce nomadic raids with the much larger larvae turning to pupae, he again removed the mother queen along with a few close attendant workers. He found her opposite clusters of spinning larvae. She was already beginning to swell with eggs. Kept overnight in the field station laboratory, the queen was to be returned the next morning, but her colony had already moved the bivouac and switched to the stationary phase. Schneirla waited to return her. The following day, with the colony now on the march, he set the queen down with about twenty of her workers. "Seven hours later, that evening, she [remains] at the same spot, in good condition and active," wrote Schneirla; "however, each time she shifts, some of the workers grasp at her legs and body, holding her in place. Two dozen or more dead and maimed workers lie around her. Under the conditions, we put her back in the same air vial for another trial elsewhere. At 11 P.M., three hours later, we set her down near the head of the column, in advance of the cocoon-carrying section. The result [is] the same: At dawn the queen, still active and in good condition, [is] in the midst of a small group of workers holding her firmly in place, close to where she had been released.... The workers [display] the disturbed, tense reactions of a group-restraint process we term 'sealing off.' "

A queen that is sealed off is kept hemmed in and tightly at bay. She cannot move, and no worker can taste or touch her. Schneirla realized that in only ten days many of the workers in the colony had changed their reactions to the mother queen. They would not execute her, as they would a queen from another colony, but neither would they let her carry on. Schneirla's actions had not hastened the process of rejection. Almost overnight, with the future of a colony at stake, workers could reject an older mother queen in favor of one of her daughters, sealing her off and preventing her from ever rejoining the colony.

Week Seven

Deep in the bivouac, in a feat of nature, the mother army ant queen lays close to three hundred thousand eggs through the first part of the week, her largest brood to date. The eggs are all workers—sterile, nonreproductive females. The bivouac is now crowded with the giant maturing pupae of the new queens and males, but the workers adjust, expanding the living brood chamber by adding more workers there. This large all-worker brood will be vital to the mother queen. Even more of her senior workers are about to desert her and take up with one of her daughters—a potential new queen. Her new brood, however, will all be loyalists—at least initially. The army ants were laid ten or eleven days later than the leafcutters, but they mature first, at thirty-four days. The leafcutters, which take fifty days from egg to adult, are still forming in the pupal stage and have another week until adulthood.

As the time of departure from the colony nears, the weather may soon play an important role. The leafcutters would ideally like to emerge after a rain so that the ground, once they mate, will be soft for digging and starting new colonies. However, it's best to have little or no rain during the nuptial flight itself. Army ants, on the other hand, are not so bothered by rain. The wingless queens never make a nuptial flight, and the males are numerous enough that rain cannot hinder them from flying far enough away to meet another colony with a virgin queen who wants to mate. The rain would have to be torrential to stop them.

Week Eight

As the leafcutters' pupal stage ends, the pupal virgin queen shows occasional tremors, body shaking that may indicate to her worker nursemaids that she is now almost mature. Around her, several other potential queens are also starting to shake. They are all removed to the perimeter of the garden chamber. The nursemaids and other adults gather around to lick the pupae. As they do, the pupal envelope gradually peels off, revealing the first winged virgin queen. A few minutes later some of the winged males emerge. All are adult size but pale in color and standing unsteadily on six untried legs. The first virgin queen slowly tries to walk. Leafcutters are rather long-legged ants, and the new virgin queen looks as if she's trying out stilts for the first time. She wobbles and jerks around but doesn't fall. This awkward, callow stage will last less than a week. Gradually, she will change to the dark reddish brown colors of an adult, and her walking will improve. Then she will be ready to leave the nest.

As the week ends, the new leafcutter adults—virgin queens and males—journey up through the long, dark tunnels of the nest. It is sometimes difficult to squeeze through certain tunnels or around bends with their wings. When they are hungry, they step inside a food chamber for some fresh kohlrabi fungus.

In one of the last chambers, the virgin queen makes a crucial stop for a piece of fungus that she must not eat but must keep in the pocket under her tongue. This is the starter fungus that she must guard with her life. Without it, she cannot found a colony.

The virgin queen emerges from the dark tunnel only to meet the leafcutter scout, coming home from a hard day's foraging. In recent months the scout has been showing signs of aging, with two of her legs wobbling and a large scar on her rear left leg. The tall, robust, winged queen beside her looks shiny, new, a symbol of freedom that offers a real chance for the future of the colony and for new generations of leafcutters. It is a chance meeting. The scout walks up to the queen and raises herself as high as she can on her rear legs to brush her antennae. At once, the scout senses the young queen's mission.

And then without warning comes the ominous sound of the phorid flies—the flies that can nearly make an ant go berserk even if they don't manage to parasitize it. If they do parasitize, however, it means the humiliation of playing host to the egg of a fly. But it is not just embarrassment; ultimately, it is death.

The flies present a clinic on how to drive an ant crazy. The scout first rears back on her legs, then lunges after them, leaping several inches into the air—with no success. It is the first time the virgin queen has encountered the flies. The scout releases a little alarm signal to alert her to the danger.

The scout's pheromone is a simple warning about the phorid flies. The scout has long experience avoiding them. The virgin queen is oblivious, busy trying to orient to the outside world for the first time, trying to unfold all four of her wings and embark on her nuptial flight—trying to begin her new life.

It is the first time the scout has ever directly exchanged pheromones with a queen. Most of the leafcutter colony would never meet a queen—their sister or even their own mother. The leafcutter scout had never actually smelled, licked, or laid eyes on her own queen mother. Yet the scout's vital messages—news of new leaf sources and her far-flung pheromone trails—had sometimes been passed through the colony, even as far as the mother queen. Her first and last physical contact with her mother had been the instant of her appearance as an egg.

The scout's warning about the phorid flies comes too late for the virgin queen to move out of the way. One particularly persistent fly alights on the back of the queen's head, out of reach of her mandibles. As the fly grabs hold and gets ready to lay her egg just beneath the integument, the queen leaks some alarm pheromone and tries to stretch her wings and fly away. Once the fly lays her egg, the queen will slowly die from the parasitic invasion.

But the scout, mandibles open, seizes the moment. She attacks the fly with her mandibles and slices the beast open, piercing its left compound eye with a rapier thrust—all without touching the queen. The dark discharge from the phorid fly oozes on the queen's back and wings. As the scout sets about to lick her clean, another phorid fly grabs hold of the scout and this time swiftly does its business, depositing the tiny, fearsome egg within the ant's brain case. A spot of whitish blood oozes form the scout's head wound. Mission complete, the fly lifts off. The leafcutter scout dispenses the obligatory full dose of alarm pheromones, but it is too late. As the summoned leafcutter soldiers emerge, towering over the scout, they, too, lunge at the fly, leaping to try to pull it down—but the soldiers are unsuccessful. Dizzy from the bite and her injection, the scout tries to stand on shaky legs. The virgin queen goes over and waves her antennae over the scout, perhaps a sort of thank-you, though none is needed in the world of the ants.

The scout has in effect surrendered her life for the virgin queen. She would have done the same thing for any colony member. The scout may live a few more days or even weeks, but her productive life for the colony is mostly over.

What motivates the scout or any other worker in an ant colony to such acts of selflessness? What is the advantage an individual ant gains in putting all her energy into the selfless tasks of a colony, from leafcutting to caring for the brood to fighting as a soldier—without ever having a chance to mate and produce offspring of her own? The answer lies at the heart of what really makes an ant colony tick as well as what makes it so successful. It is the secret of the superorganism. The ants' success lies in the evolutionary concept of kin selection. Kin selection occurs when members of a species work for the survival of relatives—genes by descent from a common ancestor rather than for one's own offspring, which is called "individual selection." Kin selection can be seen in many social species, including humans, though individual selection is the main factor in evolution. But with ants and other social insects, there is a novel twist; they take it a step further. Ants, bees, and wasps have sterile worker castes pledged to the rearing of sisters. At first, this fact unsettled Charles Darwin. "Social insects," he wrote, present "one special difficulty, which at first appeared to me insuperable, and actually fatal to my whole theory." How, he wondered, could the worker castes of bees or ants have evolved if they are sterile and leave no offspring? Darwin solved the problem by devising the idea of "kin selection." Although the workers do not reproduce, their seemingly altruistic actions preserve their fertile relatives who perpetuate the species. Thus, every ant colony is a close genetic sisterhood in which female workers derive genetic profit by helping their sisters—even more so than with human or most other animals.

Kin selection operates to such an extent in the social Hymenoptera because of their peculiar haplodiploid reproductive system. Haplodiploidy makes sisters more closely related to each other than they are to their own parents or offspring. This is how it works: Since male progeny come from unfertilized eggs, they are haploid; that is, they inherit one of the mother's two sets of chromosomes and nothing from the father. Female progeny, on the other hand, originating from fertilized eggs, receive the usual two sets (diploid)—one of the mother's two sets and the father's one set. The key is that one set—the half of the genes that daughters get from their haploid father. It is identical to every other set every other female gets, and thus the sisters must be at an absolute minimum of 50 percent related to each other. The genes from their mother, depending on which of the two sets of chromosomes is inherited, may or may not be the same. On average, based on chance, sisters share 75 percent of their genes with each other—more than the 50 percent they share with either parent. Thus sister ants in a colony can best ensure the survival of their own genes by helping each other. Males, meanwhile, with only one of the mother's two sets and nothng from a father, share only 25 percent of their genes with their sisters, on average, and predictably contribute less labor than females to the colony welfare; their purpose is reproduction through a single mating, after which they die.

In the evolutionary history of highly social insects about 150 million years ago, there is probably no more important event than the acquisition, or adaptation, of the sex-determining trait of haplodiploidy. This strange adaptation—the cause is unknown—accidentally encouraged these insects to evolve into highly social colonies and led to their "hard-core altruism."

Yet the picture in science is rarely neat. All hymenopterans are haplodiploid, yet the majority of hymenopteran species are solitary, not social. All the ants and many of the bees are social, but there are numerous nonsocial wasps. Also, haplodiploidy is not the only route to advanced social behavior. Termites, certain beetles, certain thrips, even a number of spiders, naked mole rats, as well as wolves, whales, and humans are highly social, but none are haplodiploid. The haplodiploid factor can only be part of the story of social life.

Although Darwin had suggested that kin selection might account for the sterile castes of ants, it was not until 1964 that British biologist William D. Hamilton recognized the reason was haplodiploidy. He developed a theory for kin selection and the genetic evolution of altruism to explain how it all works. Since then there have been refinements—notably by American sociobiologist Robert L. Trivers.

Hamilton's elegant explanation, championed by E. O. Wilson, among others, and refined by Trivers, has been the single most important discovery about social hymenopterans in this century. Kin selection theory explains altruism in social insects and accounts for their predisposition toward higher social life. And the ratios of females to males in the real world of ants largely fit the theory. "Somehow," writes Wilson, "ant workers manage to obey the expectations of kin selection worked out in the heads of two zoologists."

The leafcutter scout wanders off in something of a daze, marking time until her death. The flies have gone for the moment. The army ants are quiet, still a few days away from the emergence of the first callow queens, followed by the males and the start of the next nomadic phase. There is a hint of more rain in the air, but it seems distant, on the other side of the mountains toward the Pacific. As a big moon rises, the scout gazes at a last vision of so many leafcutter queens and males emerging from the colony, all perched on the mounds and at the edges of the craters of home, preparing for takeoff. A splendid sight. Then the first virgin queen, the one she had helped, lifts her wings and leaps—whooshing into the green-black of the moonlit jungle. She disappears for a few minutes and then, silhouetted in the moon, she appears, a winged ant, looking like a mythical symbol of hope, moonlight blazing through her scaly wings.

She hovers as if waiting for the males who may follow her in flight before venturing off in search of females from other colonies. They are still getting ready to take off. And then she is gone, seemingly for good, rising above the canopy, flying up, up, up.

Chapter Six

The Flood

What reason, laboring like a careful ant, with difficulty scrapes together, the wind of accident collects in one short moment.

—*Friedrich von Schiller,*

THE CONSPIRACY OF FIESCO, *1783*

Light,
death,
water,
sun,
thunder,
things that flee,
insects
that burn and die, consumed
in their own little gold lives,
torrid summer and its basket
of countless red fruit,
time
with its tresses—
everything is food that falls
into the ancient, green mouth
of the devouring jungle.

—*Pablo Neruda,*

ISLA NEGRA, *1973*

Ants Adrift. The little fire ants, tending their aphid cattle, ride their log craft down the flooded river to the open sea.

CLEARING the canopy, the virgin leafcutter queen beholds the full, starry heavens for the first time in her life. It is also the first opportunity she has had to glimpse the great expanse of jungle. A biologist would give a lot for this perspective. Down below, the canopy leaves and flowers are bathed in moonlight. From this bird's-eye view, the canopy looks like a blanket of silvery green that extends for several miles in every direction just beneath her. In area, La Selva is 5.9 square miles, roughly 3 miles long by 2 miles wide, or 3,795 acres—more than a universe to most ants. But as vast as the jungle is to an ant colony on the ground, from the air the dimensions shrink. The borders that separate the mostly primary forest of La Selva from the tamed outside world of ranches, farms, and second-growth woods become all too evident. For now, she alternately tenses her wings and floats on the dense air currents above the canopy-top carpet. Beneath this carpet, dozens of army ant colonies work or rest, depending on their shift, and other ants and insects go about their business, while several thousand other colonies of leafcutters in and around La Selva dispatch potential virgin queens and males to the sky. But where are they? In the dim light and far from the earthy smells of other ants, this virgin leafcutter queen finds herself alone, flying solo across the night sky.

For precious minutes the virgin leafcutter queen has the feel of the wind, the sensation of flying, the smell of the Caribbean on her antennae. To a normally earthbound ant, this is exhilaration. Dark clouds are looming, however, stacking up along the high cordillera that divides the Costa Rican Pacific from its Atlantic side. The summit of Volcán Barva, about ten thousand feet high, is completely obscured now. The clouds blow across the moon. The silver light flickers and disappears. She suddenly senses the dampness in the wind. Seconds later the first raindrops splash on her wings. Then a streak of lightning, and in that flash she glimpses hundreds of other flying ants—a few virgin queens but mostly big-eyed males, all intent on the mating game, intent on *her.* They were around her all the time. She also glimpses an insectivorous sac-wing bat as it narrowly misses devouring her. Like pellets, raindrops beat against her wings and all along her body. Thunder rattles the nighttime jungle world, arousing birds, and monkeys, all of whom begin to chatter down below.

At first the rain is merely a distraction from the business at hand—preparing to mate with the best flying males. But soon the rain begins to beat down, literally to pound, turning her flight into the most dangerous moments of her young life. Rain, which makes the ground soft, followed by clearing are prerequisites to embarking on mating flights, but more rain is the last thing she needs now. Forty minutes into her nuptial flight, she knows—in the way that ants know—that she must abort. There are only two things that will cause leafcutter nuptial flights to be canceled. One is a sudden drop in temperature; the other is persistent, heavy rain. Banking to the right, then to the left, slicing through the wind and trying to dodge curtains of rain, she plunges down, orienting by sight, as much as she can manage to see, to the patch of forest she just left. She passes other queens and males just venturing out—innocents unaware. Some few turn back. Most continue flying to face certain death. Even without the rain, only about one in two thousand of all virgin queens survive to found a successful colony.

Once she slips beneath the cover of trees, the rain turns to a fine spray. She lands on a tank bromeliad just beneath the canopy and almost slips into the foot-wide cistern, alive with water insects and frogs enjoying the rain. These bromeliads have spiky, fleshy leaves that grow in such a way that they form a basin to catch and hold water. The tanks become miniature ecosystems, attracting insects, birds, and other small animals whose droppings are used by the bromeliad. Bromeliads are part of the dense, lush foliage in the upper levels of rain forests that absorbs most of the moisture, soaking it up for a nonrainy day. The canopy at La Selva intercepts about 15 percent of all rainfall. On any given day, the first eighth of an inch of rain never gets past the top layer of leaves and flowers in the canopy. Only when close to a quarter inch or more of rain falls does water begin to drip through the lianas and off the leaves and down the trunks, some of which falls in a fine spray on the ground. Even so, as much as 47 percent of all the moisture at La Selva eventually returns to the sky as evaporation. The danger comes when the rain is both heavy and continuous. A recently logged forest will suffer flooding almost instantly. A second growth forest sustains the rain a little longer. A largely primitive forest such as La Selva is the best hedge against damaging floods. But with more and more of the forest around La Selva cut down, especially upstream along the Río Puerto Viejo and its tributaries, even La Selva has lost some of its powers of absorption.

And as the rain at La Selva and upstream falls faster, heavier, with no signs of abating, the sponge is becoming saturated. La Selva should still be late in the dry season, but the heavy rains have come early this year. La Selva has either two or four seasons, depending on which researchers you talk to and how long their memories are. The seasons here are nothing like the traditional temperate zone's spring, summer, autumn, and winter. Instead there is the four- to five-month dry season from January to May, a three- to four-month wet season from May to September; a one-month "somewhat dry" season in September or October, and then a two- to three-month wet season from October to January. But some researchers dismiss the one-month dry season. Indeed, in some years it doesn't occur at all, and it's certainly less pronounced than the long dry season. In their view there are only two seasons, dry and wet, which is typical of certain tropical and subtropical climates. This year at La Selva, after a short dry season, it is wet, wet, and more wet.

As the rain penetrates the waterlogged canopy, the virgin queen flies off again, searching for her ground-based home. Like a pilot eager to land in inclement weather, she finds that the ground comes up faster than she imagined.

She crash-lands. Stunned by the impact, she lies there collecting rain in the depressions on her back and in the spiracles of her abdomen.

A few minutes pass. The rain runs down her antennae and drips off her mandibles, forming a puddle beneath her head. She is soaked but alive. Securing her bearings once again as an earthbound ant, she struggles to right herself. Her two forelegs slip in the red mud. Her virgin flight has come to a humiliating end. Will she die without passing on her genes? The drive for survival sustains her.

One difficult step at a time, she moves toward the colony. Near the entrance she meets a soldier on patrol who waves his antennae around and around and finally recognizes her. Right away a worker comes forward from inside the nest as if to offer help. She limps back down into her home—the dark, fragrant nest she had left barely an hour earlier, never expecting to return. Her nuptial flight is rained out, but she has survived to fly another day—one of the fortunate few. She meets another virgin queen inside, and they touch antennae. This virgin queen never left the nest! And later she meets a few dozen other winged sisters and brothers who waited. But most of their sister virgin queens and the males from their colony are lost. There is little likelihood they will ever find their way back home.

THE SWARM RAIDER army ant superorganism is running a little late in its efforts to reproduce before the rainy season. Most new army ant colonies start to form early in the dry season, and the process is complete before the rains begin. But army ants do not postpone breeding due to rain. The rain is not so threatening to army ants compared to leafcutters and many other ants because they do not have a conventional nuptial flight. Only the males fly, and they are quite numerous and largely expendable—as long as a few manage to survive and locate colonies with virgin queens that will accept them.

Most new colonies of ants are founded after the winged males and females from various colonies depart on nuptial flights, meet each other, and seconds later mate one or more times in midair. The female then descends to the ground, rakes off her wings, and digs a hole or seals herself off to lay her first eggs. Typically, this happens after a colony has grown for a few years. The queen cares for her first brood until the workers can take over. Then she becomes mainly an egg-laying machine. Leafcutter and other gardening ants follow this general pattern, as do certain other ants.

But army ants are different. The production of males and new queens by army ants was shrouded in mystery until the 1920s and 1930s. In 1921, in Guyana (then called British Guiana), Wheeler found the first two callow queens among hundreds of male cocoons in a swarm raider bivouac. These queens had no wings, of course, and they weren't about to fly off from the colony. But, though wingless, they looked so different from the workers that Wheeler and every army ant researcher since could not help but be struck by the long evolutionary road they must have taken. And strangely, the males, when they hatched, actually resembled the queens—except that they did have wings. Indeed, army ant queens and winged males had been seen before Wheeler, but they were thought to be different species from the workers. If they look odd, the way they start new colonies, compared to other ants, is unusual, too.

Something strange happens in the army ant superorganism when the young queens begin to emerge from their silk cocoons inside the bivouac. The superorganism begins to divide. Fission was Wheeler's favorite hypothesis as early as 1923, but Schneirla proved it with his exhaustive field studies, mainly on Barro Colorado Island, Panama, beginning in 1932. It is only following the raising of a royal brood that division happens. The final stages leading to division begin when the workers caring for the royal brood develop an enduring attachment for them. These larvae soon become adept in the chemistry of charming other ants. Based on their attraction skills, they enlist the support of their fans, and soon they have an entourage, consisting of their sister workers stolen from their mother. The larval queens that attract workers early and amass the most supporters have the best chance of becoming full-fledged breeding queens with their own colonies.

At dawn, as the royal brood of three thousand males and six queens nears maturity, the swarm raider workers emerge from the bivouac suspended from the aboveground roots of a tree. It is raining steadily but the ants are oblivious. Two strong raiding systems appear—one to the north, the other to the south. These are the last raids of the stationary phase, soon to be followed by the migratory phase when the bivouac will be moved nightly. More important, these are also the last raids of this existing colony.

Deep in the protective bivouac, the first callow queen turns into a full adult. Now the beauty contest begins. First dozens of workers, then hundreds surround the new virgin queen, licking her. These "turncoats" are so taken by the smell of the pale, golden yellow virgin that they switch allegiance from the colony queen, their darkly pigmented mother, to the new, fairer queen, their sister. Other workers, still loyal to the mother queen, stay with her.

At a fortuitous moment, with the virgin queen's star on the rise, the mother queen and her supporters storm out of the bivouac on the southern trail, preparing to run as far and as fast as they can to form a bivouac and hold it together. If she can run the length of the trail before any of her queen daughters, she will be one of the two queens to form a new colony. Now five years old, the known upper age limit for army ant queens, this queen has kept her workers attracted to her for several years during the production of the annual royal broods—ever since she was a virgin queen. But this time she doesn't get very far. About ten feet along the trail she meets a phalanx of younger workers who are deeply agitated at her presence. They move their antennae all over her but are no longer interested in licking her. They hold onto her, pushing then pulling her away. She is knocked off the trail along with her loyalist workers. Still the younger workers keep pushing her. Seconds later she breaks away, trying to run back to resume her passage down the trail. Her tarsal claws dig into the mud as she fights to regain her position. She makes it to the trail and one more time inhales the deep perfume of the swarm raiders in action, but she is blocked again and again and finally herded off. She is cornered by agitated workers who proceed to "seal her off," as Schneirla terms it, from the rest of the colony and from the trail. Some of the older, darker workers surrounding her are killed, but she and her now diminishing band of close loyalists are left unharmed.

A few hours later, with the north and south trails still strong, the young virgin queen, accompanied by her retinue, charges off on the southern trail. This is the trail nearest to her position in the bivouac—the trail established for her by workers she had begun to attract in the bivouac. As she appears, a deep perfume odor fills the air and accompanies her and her retinue as they advance steadily along the trail, gathering up more and more workers. She moves out past her mother queen to assume her new status as founder of the southern daughter colony, some twenty to thirty feet farther along the trail. She waits then, protected from the rain by some of her workers, while others gather food for the new colony.

Within the hour a second virgin queen takes the north exit from the bivouac and, mobbed by her sympathizers, proceeds more than fifty feet along the northern trail. Her perfume is different, though just as aromatic as her sister's. She is in turn followed by four more of her virgin queen sisters, each with a small coterie of workers and less attractive smells; none are serious contenders. They advance only a few feet along the northern trail. Minutes after they emerge from the bivouac, the four are sealed off. The workers supporting these queens mostly defect or are isolated with their hapless would-be queen. The second victorious virgin queen now effectively establishes the northern daughter colony.

The two successful virgin queens have left home, albeit the often mobile home of their youth, for good. But they have also taken a piece of it with them—the workers and a new worker brood, as well as the heavy males in their bulging cocoons—all of which once belonged to their mother. For the last time they have enjoyed the rich warmth of their mother's bivouac. Now there is just the new, powerfully attractive smell of the virgin queens and the individual smells of their colonies. Now they must turn to essential tasks: to hatch their own males and get them out into the world and to enable a foreign male to enter into the new colony to mate and ensure its future.

The males hatch later in the day. Their emergence makes the workers go into overdrive. They cluster around the males, eating the pupal tissues still attached to the callow males. From the start the workers are deeply attracted to their brothers, who are growing stronger by day, yet within four or five days, as their brothers depart on solo nuptial flights, these workers will need to shake their allegiance and transfer it to a foreign male who will be their own colony's ticket to the future.

By midnight, the rain having abated for an hour, only a thin two-way column connects the north and south bivouacs where the new colonies have taken up residence. This is the last link between the dividing colonies. Workers unsure of their allegiance go back and forth between the two. The mother queen still has a few loyalists camped out with her a few feet up in a balsa tree off the main trails, but she is largely forgotten, superseded by her daughters. Even though she had a fast start on her daughters and had left the bivouac first with her loyalists, she is unable to establish a colony. Over the past five years she maintained her status through several colony divisions. Usually, the queen mother keeps half of her colony when the colony splits, but sooner or later she loses everything and is abandoned when it is no longer in the best interests of the superorganism for her to continue as queen mother. She is simply no longer attractive—or not attractive enough—to the workers.

The first two or three virgin army ant queens to emerge are the leading candidates to form new daughter colonies. Of these, usually only the two most vigorous queens survive to carry on.

In an effort to understand more about the mechanisms of colony division and what happens to the mother queen in these circumstances, Schneirla watched many army ant colonies and studied them in detail, capturing queens on various occasions and performing experiments on them. One swarm raider mother queen that he captured after a colony had split "turned out to be a real beldame, very darkly pigmented, with many ventral scars, and an elongated and sagging gaster." These signs of age partly explained her rejection by so many of her workers. Yet she still retained supporters. He also removed a few of her large cluster of dark older workers, as well as random workers from the colony she was in the process of losing. He took them all into his makeshift lab on Barro Colorado Island and put the random workers and the old queen together. Immediately, they started nipping her legs and antennae, and pushing and pulling her around. Next he tried the dark older workers from her entourage—her loyalists. Their behavior was just the opposite. They surrounded her in a protective, supportive way, licking her gaster and mandibles. Although this queen and her supporters had been the first to leave the bivouac to try to establish one of the two newly divided colonies, Schneirla realized that success depended not only on attracting a good number of workers but in not being offensive to too many others. This queen's day had come and gone. When Schneirla put the queen and her entourage back into place off the trail, resistance to her continued, and eventually she was abandoned by her last workers and left to die. Though she still retained some of her old attraction, it was not sufficient to command the allegiance of enough workers to take another colony through another year.

FOR TWO DAYS, as it rains off and on, and the rainy season becomes a fact of life, the virgin army ant queens with their workers begin to establish their independence. Their raids are rather weak and their new life still tenuous. On the morning of the second day, the new virgin colonies are still connected by a two-ant trail of workers moving in either direction. The traffic, much less than the previous day and growing thinner by the hour, is composed of confused workers still unsure about their allegiance. By midmorning it disappears. Except for a few workers still attending neglected queens, all have gone with one or the other.

The males are fairly evenly split between the northern and southern bivouacs —about fifteen hundred in each colony. At first very weak, the males are held in strands of workers that form the walls of the bivouacs. The males are treated well by their worker sisters. Also hanging in the walls like grapes in an arbor are bits of booty that the workers have brought back for the males. To feed, the males have only to put their mouths up to the juicy morsels and suck out the liquid protein. It is unknown whether the males eat much solid food or to what extent they depend on their own tissues to get them through their short but intense three-week lives, but at least the hanging booty is there if they want it.

During the first couple of days of the new colony, the males accompany their sisters on the migratory raids. As the males walk along, they keep closely in line. Each is trailed by his own retinue of adoring sisters, some of whom run with their mouth parts pressed tightly against the male's thorax or gaster. Sometimes the smaller sisters, called submajor workers, partly ride on the gaster, grabbing hold of the male's wings in their mandibles. For some reason the males are even more attractive to their sisters when they're out running on the trail. First one sister will grab hold of a male's gaster, then another gets a chance to ride, clinging to the wings. As the march progresses, the excitement over the males increases; however, their attention to their brothers does not much interrupt the workers' work. They still hunt by day as part of the raiding parties and by night carry the brood and booty, as needed, when everyone moves to a new bivouac site.

Three days after the colony split in two and the males hatched, the males are feeling stronger. On the fourth day they begin to break loose, running off the trail, making little skipping runs over the leaves. They also mount the vines and other sloping surfaces, pressing the hairy tips of their gasters against them in simulated sexual intercourse. Their worker sisters respond by clinging harder to them. The result looks like a Roman circus. As the males in the southern colony bolt off the trail, hopping short distances with their wings, which are now starting to work, their sisters race after them. When the females catch hold, their grip on their brothers' wings is tenacious. With hundreds of males and thousands of female workers trailing after them, the scene is chaotic—nothing at all like the typical, well-organized swarm raids.

On night four, shortly after a new bivouac is established, the males begin to rush out. They push and shove, clearly in a hurry. The workers try to restrain them but find it increasingly difficult. One by one, like a storm trooper, each suddenly takes off, never mind his sisters and the rain. He steers boldly, if a bit erratically, in a steep ascent, hoping to clear the upper canopy where the wind will take over and blow him to a new destination perhaps a thousand feet away. Each flies for one purpose and one purpose only: to find a friendly army ant colony with a virgin queen whose workers will accept a foreign-smelling male and let him acquire the colony's smell so that he can mate with its virgin queen. He has no more than three weeks to accomplish his life's purpose. As robust and well equipped as they seem on land, the males are rather defenseless in the air. A few may be washed away in the rain, but most are more likely to end up as juicy flying morsels, snapped up by flying insects and birds. Of those males who succeed in getting through the early stages of what Schneirla calls the obstacle course, few manage to find a foster colony. It is the rare male who lives to achieve his only goal and function in life: to mate with a queen.

The effect of the sisters' behavior toward their brothers is to help them along by delaying their flight until they are at peak strength. Also, because they fly off over a several-day period after the mother colony has split in two and the two daughter colonies are on the march during their first migratory phase, their distribution is extended. As Schneirla puts it, the superorganism effectively seeds a large area with "winged sperm carriers."

Colony reproduction in army ants follows the growth and reproductive schedule characteristic of flatworms, sea anemones, and certain other animals that use fission to reproduce. In studies mainly in Panama by Nigel Franks, an ecologist and myrmecologist at the University of Bath, England, tests of the predictions from the mathematical models for the growth of such organisms reveal that the swarm raider superorganism fits the pattern. The superorganism grows to a size at which its efficiency starts to decrease. By splitting in two nearly equal parts at this optimum point, the daughter colonies carry on at peak efficiency and growth. Each new colony gets not only half the worker force but half of the males and half of the new worker larvae from the original queen. The splitting also typically occurs during a period of the year when prey is thin on the ground—the dry season. In seasons of slow growth, smaller-size organisms are more flexible and efficient at gathering food. Two superorganisms can cover a larger area than one. In this way, reproduction by fission maximizes the swarm raider superorganism's ability to survive and carry on.

As the last male flies off on day six, the rain clears enough for the two new virgin colonies of army ants to launch massive swarm raids. The northern colony heads for the darkest part of the forest, while the southern marches toward the swelling Río Puerto Viejo. These raids are much larger than usual and have two base or trunk columns. As all the workers depart with either of the two virgin queens, the old mother queen and the other four virgin queens, still sealed off with their dwindling workers, are out of luck. They are not killed, but they are abandoned to die. The workers have made their "decision," cast their fate with the two strongest, healthiest, most *attractive* queens. In the days and weeks to come, how well they have chosen will become clear.

ONLY A LITTLE downstream, along the soggy banks of the Río Puerto Viejo, the leafcutter superorganism is under growing threat. The architecture of the mounds and craters of the nest mostly funnel the rain off the roof, but as the rains increase, the numerous scattered entrance holes begin to funnel water down inside. The holes become more dangerous and must be plugged up. The lengthy tunnels through the nest, as crucial as they are to daily life—including all travel, communication, and food and brood transport within the colony—become gushing water pipes directing water to the heart of the colony, sweeping away and drowning ants in its path. Fortunately, it is difficult to flush out the delicate gardens and larval nurseries from a leafcutter nest because the connecting tunnels enter and leave each chamber at the base and the gardens and brood grow up inside each chamber. Exterminators who have tried to force poison gas or liquids into the chambers to kill the ants have mostly been frustrated in their attempts. One researcher, however, was successful in pouring wet cement down the nests, then excavating the dirt around to determine the architecture. But not all the chambers would fill up with cement. When the water table rises, though, the chambers will start to fill up no matter what. That is starting to happen already. First to fill with water are often the adult and empty chambers on the periphery, then the brood chambers, and finally the centrally located chambers, where the queen lies, the most vulnerable yet essential colony member of all. She is larger than an army ant queen, and her own bulk and weight make her slow moving. But the first threat is water pouring down through the tunnels, and the most vulnerable to this is the new virgin leafcutter queen who managed to return to the nest and now waits just inside one of the entryways for a chance to fly again. She must be protected.

After one of the west tunnels starts to become waterlogged, hundreds of workers struggle out of the nest and through the rain to plug up the remaining entrance holes in the colony roof. They collect and push dirt into the path of the water, trying to seal the colony. As an exercise, it is three steps forward and two back because most of the dirt carried to each entryway by an ant is instantly washed away. Eventually, by degrees, the task is accomplished.

A few minutes later, however, from inside the colony, a soldier and several workers feel steady vibrations coming through the ground, possibly from outside the nest. The vibrations—actually short, transient clicks—cause one worker to emit a pheromone that says, "Let's cluster together and dig." The soldier steps aside, and nine workers together excavate one of the tunnel entrances, emerging once again into the rain. They begin running circles around the outside area, trying to pick up the vibration again. It can scarcely be felt above the rain's persistent pounding on the bare earth. These ant vibrations—called stridulations—are made by various gardening ants, but leafcutters have the highest frequencies. At close range they can be heard as a faint squeaking by the human ear. The ants apparently cannot hear the sounds at all; they only feel them through the earth. The sounds are produced by an ant moving its gaster up and down in such a way as to cause a filelike part of its abdominal segment to rub against another part like a scraper. Stridulations are one of the communication techniques used by various ants such as leafcutters. Stridulating in leafcutter ants is a signal that means "I am trapped underground"—a message that once received is quickly acted upon.

Orienting to the vibrations, the leafcutter workers finally locate the source. A few wings are protruding through the mud. They dig down and slowly uncover three dirty, straggly winged queens that never made it back to the nest. They are half-dead. The gaster of one is just able to move up and down to cause the stridulations. All three queens are rescued and carried back to the nest; two die in transit. They are still given royal treatment, however, because it will be one day before the death smells begin to emanate and they are carried off to the rubbish heap. The third winged virgin queen has taken in some water but appears to be recovering. The workers gather around to lick her wings and abdomen clean. For a day she looks healthy, waiting with a sister virgin queen to fly again, but then she suffers a relapse and dies. The next day when she is carried away, the virgin queen who aborted her flight and got back to the nest takes up her vigil near the door. She wants to try to mate and start a new colony. She wants to make her nuptial flight again. Still, she must wait.

Some species of ants that often encounter high-water conditions have evolved special coping strategies. In the early 1980s, E. O. Wilson discovered the flood evacuation response in a *Pheidole cephalica* ant colony he had brought back to his Harvard lab from La Selva. Wilson noticed that when the minor or smaller workers of the colony find a single drop of water at their nest entrance, they respond by "making alarm runs through the nest." They vibrate their body at other members of the colony and lay an odor trail away from the drop of water. In less than thirty seconds, one or two workers can get a nest cleared; all workers, brood, and queen depart by the "back door" or out secondary entrances. The reaction to so little water impressed Wilson, but he also marveled at the neat complexity of ant communication. In fact, the flood evacuation response is the reverse of recruiting fellow ants to a food source outside the entrance and along a trail. The unique chemical message—together with the visual signal that apparently enhances the effect of the chemical—tells the other ants that this is an emergency situation. To confirm this behavior in the wild, Wilson took his *Pheidole* ants back with him to Costa Rica the following year and put them through their paces. The experiment worked just as well in the field. Then he brought them back again to Harvard, making these *Pheidole* ants some of the most well traveled ants in the world.

There are many other strategies for dealing with excess water. In the floodplains of the southern United States as well as in the savannahs of northern South America, fire ants respond to rising floodwaters by forming living rafts. As the waters flood their underground nests, according to Wilson, the ants move up to ground level, forming large masses that can float. The superorganism drifts until it can attach itself to a bush or grass stem that sticks up above the water. Some of the workers may drown, but the crucial brood and queens survive in the center of the mass. When the water recedes, the ants return to the soil.

Other species of ants don't bother moving but can survive long periods underwater. Submerged *Formica* queens and workers can live two weeks or longer in an anesthetized state during which their oxygen consumption falls to between 5 percent and 20 percent of their usual resting rate. Ant species that live close to water or in bogs subject to flooding seem to have evolved an ability to remain submerged for long periods if necessary and to extract oxygen from water.

Leafcutters are not known to have a tolerance for water or an ability to withstand flooding, apart from the architecture of their nests, which naturally drains a certain amount of water. Furthermore, they don't move their nests as readily as other ants. Part of this is because they invest so much time and energy building large elaborate nests. For gardening ants it's not just moving the queen and brood but the whole farm as well. This means chopping up the delicate fungus gardens and carrying them piece by piece, all the while protecting them from alien fungi, bacteria, and insects that might invade the morsels and harm the fungus that provides the ants' sustenance. In the drier parts of Costa Rica, a leafcutter nest might stay in one site for the life of a successful queen, up to two decades or more, but in the Amazon basin, some nests have to be moved every year or even twice a year due to flooding.

The overriding instinct of the leafcutter superorganism at La Selva is to protect and repair what it has built rather than evacuate. The workers continue to plug up the entrance holes with dirt, mud, and leaves, as well as with bodies of drowned ants. For the moment the dike is holding.

LATE ON DAY SIX, the rain nearly doubles in force and becomes torrential. The northern swarm raider colony locates a secure log on high ground and sets up a temporary bivouac suspended partly from a tree branch but situated mostly under cover inside the hollow tree. Meantime, the southern colony becomes stranded as it scrambles for cover, and a section of the Río Puerto Viejo splits off and rushes on either side of it. There are many casualties among the nearly blind workers, hundreds of whom are swept downriver, but enough of them are able to move to a high branch over the river. Ant by ant they attach themselves to a liana wrapped around the branch and interlock their legs, forming a living bridge down to the base of the trunk. They completely surround the queen and keep her warm and relatively dry.

Farther down the riverbank, keenly aware of the southern virgin colony, a foreign army ant male bides his time. He just blew in from the coast. It had been a rough ride, but now he smells his reward. The next time the southern workers emerge on a swarm raid, he will try to slip into their midst and acquire their smell and, with luck, access to their queen. For now he waits, the rain dripping from his wings.

• • •

FROM INSIDE THE leafcutter colony, more stridulations through the ground renew the call to action. One potential queen is rescued, increasing the colony's chances of getting its genes into the next generation, but the excavation proves costly for the colony. The uncovered hole lets in a lot of rain, and now the workers have trouble plugging it again. Meantime, the water table is slowly rising. The Río Puerto Viejo rises ten feet in an hour. Through the morning it rises thirty feet. By early evening, it has increased by almost forty feet, and the rain continues to pour.

Rain in the tropics can deposit a great deal of water in a short period of time. This is the characteristic tropical thunderstorm. Sometimes the so-called rainy season means only an hour or two of rain in the afternoon. After the rain, the humidity may be relieved somewhat until it starts to build up again later in the day. But now the rain is more persistent, and the overflowing river is threatening trees, despite the special buttressed trunks, an adaptation that allows the trees to survive for some weeks or months in standing water and also adds support to the shallow roots. Six days of off-and-on rain now threaten to overwhelm the rain forest canopy. The tank bromeliads are all spilling cascades of rainy waterfalls. The weight of water held in the canopy makes the trees creak and moan. Soon comes the ripping sound, followed by the thud as a tree hits the ground. Another falls at the edge of a clearing, then another and another—a domino effect. Trees hold the rain forest together. Every tree lost means that there is one less to absorb the rain and protect the earth. In former days the lushness of the Atlantic lowland tropical rain forest extended in a continuous band from the Yucatán of Mexico to the Amazon. Today, trees are increasingly cut down for ranches, roads, and farms. Costa Rica, with its extensive park system, offers more protection for its rain forest than most countries. Percentage-wise, it has one of the highest park-to-total-land ratios in the world: 11 percent. But it also has among the highest deforestation levels. By the mid-1990s, Costa Rica became the first Neotropical country to have virtually all its forests not in protected parks cut down. As the Central American and South American tropics become more deforested, floods will steadily increase in number and size.

One of the fallen trees, not far from the leafcutters' nest, is a large, mature cecropia that only the previous year had achieved canopy status. This was the same tree that the miniature army ants had briefly explored—only to be repulsed due to the symbiotic association with the tree's own standing army, the fierce aztec ants that aggressively drove off all comers and exploiters. But the fast-growing tree lacked structural strength. Now the death of the tree poses problems for the aztecs. For the moment, the queen remains inside with a band of workers, some of the brood, and a few head of homopteran cattle, the honeydew-secreting insects that they regularly tend for food. Many workers explore the situation outside and are washed away. Sooner or later, they will have to leave this dead tree and find a new, young cecropia to move to for food and housing.

ON THE EVENING of the seventh day of rain, the virgin leafcutter queen sticks her head out in hopes of resuming her nuptial flight. She ducks back in quickly, the water almost washing her down the tunnel. The rain begins to take on the character of a flood. Although not quite a flood in human terms—that is, not dangerous to human life or property—it is life threatening to the ants, insects, and other small creatures, the most numerous and important owners of this forest.

Ants are part of an order of insects that, with their highly developed social systems, have become masters at adapting to environmental change. But what they are about to go through challenges even their ability to respond quickly, carefully, and correctly, and thereby to survive.

The leafcutters are drowning now. It is one thing to be sitting out the rain resting in a tree, even if it's a fallen cecropia, and it's quite another to be living underground, exposed. Only one of the leafcutter virgin queens managed to return to the nest safely. A number of others who never left the nest remain below, still to try their wings. Some of these winged reproductives are caught trying to escape, trapped underground and buried alive in mud. As the water table in the nest begins to rise rapidly, the nest is overflowing. Rivers of dark brown jungle ooze—leaf pieces, fungal mycelia, and ant corpses all dipped in mud—are spewing from the blocked-over entrances and ventilation holes. The refuse pits are backing up, and now spent, moldy pellets and rotting corpses of ants are floating everywhere. The colony is quickly becoming flooded.

It's time to move house. Inside, the leafcutter workers patrolling the dry tunnels at the center of the colony assist the queen and as much of her brood as they can carry and head up the central ventilation shafts where one worker has indicated there is a clear path out of the nest, a path that avoids the rushing water. The path smells faintly of naphtha, the grassy, fruitlike aspect of the smell dissipating in the rain. But even in the rain the message is clear. Other smaller workers start dicing up the few surviving gardens and giving a piece to each worker to carry. A long, damp procession leads up through the nest and out the central shaft. The path forged by several of the leafcutter workers then leads toward higher ground—up a slippery slope and into the cover of several massive canopy trees. There is no time for ceremony as the first worker breaks new ground and clears away the first tiny piece of mud with her mandibles. The task is just beginning; it will take days to construct a few chambers and tunnels of a new underground nest and transport the colony and farm to safety. But the superorganism is committed to the move, and one and all have sprung into action.

Floods are an integral part of the annual cycle of life in many wet lowland tropical rain forests. During the alternating dry and rainy seasons of the tropics, creek beds subside and swell, rivers run clear and slow and shallow, then brown, muddy, and fast—and few animals or plants in the forest, large or small, escape the changes. Many, in fact, make use of them or time their reproductive cycles—from flowering trees to insects. Even when the floods come, it's not all negative. The numerous canopy dwellers make a feast of the luckless ground insects that desperately walk, run, or fly up to the canopy to escape rushing waters. And there are some ground species that take advantage of the corpses washing up everywhere.

Nowhere in the tropics do regular floods have more impact than in the great Amazon basin, which expels into the sea one-fifth of all the river water on Earth. During the rainy season, the floodplains of the Amazon can extend thirty to sixty miles wider than the main channel of the river. In height, the river fluctuates by at least forty feet a year, and levels of sixty-five feet can occur. When the Amazon is at full strength, the freshwater flows so far out to sea that it can be detected some two hundred miles east of the river mouth, a tenth of the way to Africa. The implications for the dispersal of various animals are profound. Most insects and animals that cannot climb trees or scurry to dry land die, but others grab hold of pieces of wood, floating seeds, or assorted flotsam, and some few survive to land on distant shores. The process of disseminating species this way was once described by the American evolutionary biologist George Gaylord Simpson as "sweepstakes dispersal." The chances of a species surviving are about equal to winning a sweepstakes.

Along the Río Puerto Viejo, however, the floods are never as extensive as the regular cycles in the Amazon. The driest weeks of the year at La Selva are from February to April, with some weeks, particularly the last week in March, often with no rain and a ten-year mean of less than half an inch. The main wet season begins promptly by the second week in May when ten times as much rain, about 5 inches, is the mean, and nearly 11 inches the ten-year record. The rainiest week in the past ten years had 13½ inches. The worst flood in recent memory at La Selva dates from 1976 when animals as well as biologists were caught unaware. Some of the researchers were stranded in the forest and had to swim over submerged bridges to get back to the station. That year, 177 inches of rain fell—below the record 223 inches in 1970 but above the twenty-eight-year average of 156 inches, or 13 feet.

The leafcutters work through the night and the next day to move most of their nest to their new home, while other leafcutters manage the new excavations. As the rooms are ready, the most valuable parts of the colony—the brood at various stages from eggs to pupae—are transferred along with the bits of cottony fungus that are attached and sometimes growing all around it. It takes more than 5 million carrying trips between the two sites. Several thousand workers and almost half of the brood are lost in various landslides and cave-ins, but the move is a success. Even the dying scout is successfully moved and, more important, the virgin queen, who with any luck will be able to fly off one cloudless night soon and mate. Almost last to go is the massive queen of the leafcutters. Six of her nurses lead her out of her chambers. Still others follow behind, taking turns licking her and attending her as she lumbers up through the tunnels. Two minor workers ride atop her gaster, licking her from above. The trip through the rain is fraught with danger. With the queen out in the open, the colony is the most vulnerable it has been since its founding. But few predators of ants can be bothered to hunt on such a day. Most are too busy trying to flee for their own lives.

As the mother queen resettles in her new chambers deep inside the new nest, the virgin queen takes up a new post at the colony entrance. Several workers keep watch over her, as if to make sure she doesn't leave too early. Walking endlessly in circles, then standing still for hours, she is not doing a rain dance but, it would almost seem, some version of a crazy sun dance.

Meantime, edging into a new back eddy of the river is a massive cecropia, which swings out toward the full stream and breaks away with a crack. Many aztecs are still aboard. The hollow connecting chambers of the trunk are quickly flooded. Those ants that escape try to hang on the trunk or on a branch, but most go spinning to their deaths. Seconds before the tree breaks, however, a tiny, precious contingent of the colony containing the queen and some core workers manages to escape. They climb aboard a nearby standing tree, find a safe hollow, and wait.

THE CECROPIA, with a few bewildered aztecs running up and down the river-washed trunk, passes spiders bobbing along on their own air-filled egg sacs and luckier ant colonies carried intact inside buoyant cacao pods. There are also various small mammals that never made it to high ground. A domestic pig, astraddle a piece of wooden fencing, rolls by. All have become part of what La Selva canopy biologist Donald Perry has called the "soup of desperate life [carrying] forlorn passengers to unknown destinies." One river passenger *not* so desperate is a small but fierce colony of little fire ants that Wilson and Brown nearly collected the previous month on their first day in the field at La Selva.

After Wilson and Brown's tinkering, the same little fire ant superorganism, or one very much like it from the same area, managed to crawl deep inside a seven-foot-long log. The little fire ants brought along their own whitefly and aphid "cattle." When the rains came, however, they lost some of their cattle, while others were brought deep inside the nest for protection. Then the log, more in a low-lying part of the forest, became surrounded by water. As the river level rose, the log stirred, and then began rocking to and fro, and the little fire ants hunkered down. Finally, the log floated free, bumping and brushing along the eroding bank before it slipped down the river, full steam ahead, as if in search of the grandest adventure of its life.

Edging through and emerging from the flotsam and jetsam of a flooded rain forest, the log begins to resemble a sailing ship with sleek lines. This seaworthy log-ship has two slender branches, actually deeply rooted epiphytes each about three feet high, that stick up above the water. Both are covered in leaves and entangled with vines. From a distance the impression is similar to a two-masted schooner with its masts slightly askew but nevertheless under sail and moving fast before the wind.

In the darkness of the closed canopy tropical rain forest, the little fire ants are almost inconspicuous—despite their golden color. The workers are minute, the superorganism modest in size. Only the sting, fiery for a creature of such small size, distinguishes them. They are kept in check by the sheer number and density of ants and other insects of the rain forest. This includes aggressive ants such as aztecs; those dominant storm troopers, the swarm-raiding army ants; as well as, if only in spatial terms, the massive leafcutter colonies, which already occupy large territories. But removed from the diverse, species-rich rain forest and given the chance to move to a species-poor environment, the character of this ant changes. It becomes an uncompromising opportunist. It travels well. The little fire ant is a so-called tramp species, a sometime stowaway in lumber, fruit, and crates, ready and eager to be transported to any preferably tropical or subtropical port. It tries to bring a few of its cattle with it, but the species is more flexible and less dependent on insect cattle than other ant species such as the aztecs. It travels so well that the little fire ant is frequently found in quarantine stations in foreign ports awaiting an exterminator to dispatch it before it enters and wreaks havoc on some new corner of the globe. But the little fire ant really flourishes on islands and other species-poor habitats. Like Genghis

Khan, the little fire ant superorganism kills the weak and destroys or otherwise displaces everything that gets in its way. From its origin, probably in tropical South America or Central America, it has spread to islands as diverse as Puerto Rico, where it is known as *albayalde* to much of the West Indies and to the Galápagos Islands in the equatorial Pacific. Its conquest may be subtler than Genghis Khan's raping and pillaging across Europe, yet on a smaller scale it is every bit as pervasive and complete. Still, it doesn't kill everything. In the Galápagos, for example, the native carpenter ants seem to do fine. When it visits most places, it slips in quietly, concentrating on marginal habitats, but within a few years it gains a solid foothold. A few years ago Brown noted that the species had spread across the South Pacific, beyond Fiji, as far as New Caledonia, just nine hundred miles short of the east coast of Australia. He published his report in an Australian entomology journal to warn the Australians that the little fire ants might soon be quietly storming the beaches. In fact, they may already be there, but in the early stages of an invasion and depending on where they land, they may still be inconspicuous. The long-term effect on Australia, with its extensive endemic fauna, could be devastating.

Riding in the main channel of the river, the schooner log with the little fire ants heads downriver through the night, and the smell of the Caribbean, the Atlantic side of Costa Rica, grows ever stronger. Various drowning creatures try to jump aboard this ark along its way to the sea. A pregnant wolf spider manages to slip on board and dive underneath the hanging shards at the end of the log—practically out of reach of the little fire ants due to the lapping river. Several ants and termites are not so fortunate. One termite king and queen are torn from their makeshift nest at one end of the log. Several bala ants—unable to climb a tree in time—pace the log, trying to avoid the little fire ant workers. The balas, with their long legs, can easily outrun the smaller ants, but there's nowhere to go on this log. After two hours of avoiding each other, one bala and four or five of the little fire ants lock mandibles and fight to the death. Two of the five little fire ant workers survive to claim victory. Shortly after the battle, a herd of whitefly and aphid cattle are marched out of their nest to feed on the leafy epiphytes. Each is followed by a little fire ant worker who will collect any honeydew—the sweet brew that helps keep fire ants highly charged and fiercely aggressive. But an hour later, just before dawn, the log bumps into a floating tree, and a few dozen aphid cattle, as well as the remaining balas and several little fire ant workers, are all knocked overboard, watery graves for all. The rest of the cattle continue to graze and be milked by the fire ants—all unfazed, though now gripping the branches and leaves much tighter with their tarsal claws. They're getting their sea legs.

As the raft reaches the mouth of the river, a leafcutter male—who has obviously left his colony a little prematurely and found no females—lands on the little fire ant colony's schooner. He is exhausted but welcome. The little fire ant majors come out, surround the huge animal, and promptly sting him to make sure he is subdued. Short of meat on their hastily provisioned ship, they welcome the fresh carcass. They dismember the big male, wings first, then the legs, followed by the tender, tasty gaster.

Lucky all the way, by dawn, the little fire ant superorganism now passes out through the estuary and spills into the open Caribbean. The ants taste the salt spray on their antennae. The waves—long, rolling, early-morning swells left over from yesterday's storm far out at sea—set the schooner pitching and rolling. The wind is up, and the currents join forces to push the little ship north by northwest, almost parallel to the land, along the coast of Nicaragua toward the Corn Islands. Fortunately for the ants, the base of the log is flat, preventing the ship from rolling over completely. And the two branchlike epiphytes covered in leaves are acting as crude sails, allowing the ship to advance slowly but steadily before the wind. The hull speed is perhaps only three knots, but this is faster than the current. The added motion does not bother the ants, but the occasional water washing the decks causes them to sequester themselves and their cattle deep inside the log. They have battened down the hatches, and now they are well on their way to Cuba, the Dominican Republic, or perhaps other islands where others of their kind have visited before. Should they miss these islands, however, they might well hit the Gulf coast of Texas or Louisiana, or perhaps veer east, where the wind and currents could take them to Africa. The world beckons. There are many places still to be conquered by the little fire ants.

BACK AT LA SELVA, as the rain abates and the river level stops rising, the southern virgin army ant colony, bivouacked above the river, dispatches a swarm raid at dawn. By mid-morning, when they are spread out along the trail, the winged swarm raider male who has been biding his time outside the nest strolls over to the path and tastes the aromatic molecules hanging in the air. He ambles along the trail for a hundred feet, then stops. The army ants are returning with some of the booty. He steps aside for a moment and waits. When they have passed, he snaps into line behind them. His task now is to obtain the scent of the colony. The only way he can do that is by hanging around the workers and insinuating his way into their lives before making any advances toward the queen.

Almost right away the workers turn around and begin to fête the strange male. His scent is working. With every lick, bump, and casual touch, the workers help to bring him into the colony and make him smell nicer for the queen. With all the jostling, his wings soon begin to feel loose. Several of the workers pull gently on them, and they drop right off—more like removing a dinner jacket than extracting a tooth. He will not need these wings anymore. He feels lighter. He is now surrounded by many more workers who are even more attracted to his winglessness, for this makes him much more like the queen.

Schneirla was the first to suggest that a male must acquire the colony's odor before he will be accepted by the queen. Then, after losing his wings, he may live for a period of time in his adoptive colony—days or even weeks—before mating. But this male wastes no time in approaching the female.

Marching back with the workers to the bivouac in the late afternoon, the male's assimilation is complete, the southern swarm raider superorganism is now about to lose its virginity. That night, as thousands of workers mill about the queen inside the bivouac, the male walks across the bodies of the workers, who seem to lift him as he goes, as if to make his step lighter. He smells his mate. She alone among all the ants in the colony is his size and has a similar physique. Even in the low, dappled light of the bivouac, both bodies glisten, their armor having been polished to a high sheen by the licking of so many attentive workers. Facing each other, the two ants exchange licks. They lick each other again and again. The male now reaches for the queen from behind with his middle and hind legs. Using his mandibles, he grasps her petiole behind her horns. He quickly inserts his sexual organs. He does it with such force that the queen's gaster appears deformed. In closely related army ant species, such as the column raiders, the male grabs the petiolar horns of the queen in his mandibles and remains entwined for two to ten hours. In one ten-hour coupling noted by Schneirla, the male was dying by the end. But this mating lasts just one hour—similar to the only published account of a swarm raider mating, observed by Carl Rettenmeyer after he had brought a pair into his lab.

The army ant workers, unlike most other ants, play an active role in the reproductive life of their superorganism. With most ants reproduction happens high in the air or far away from any workers. But the army ant workers are all around. They link legs to form the bivouac, which turns now from sometime nursery to the temperature-regulated boudoir for the queen to receive the male for mating. They witness the mating of their queen and the male, smelling the sex. But most important of all, they actually assist in choosing these two individuals who mate in their midst—the matriarch and patriarch of the new colony. In this case they chose a new queen from among their virgin queen sisters rather than keeping their mother as the matriarch. And then they, in effect, selected their sister's mate—their patriarch.

Wheeler and other entomologists often compared ant societies to human societies. Part of this was a search for a Bigger Biological Picture. Wheeler himself, who was a socialist, considered the cooperation in ant societies a good model for human society. Others, such as Caryl P. Haskins in his book *Of Ants and Men* (1945), characterized the small superorganisms of primitive ponerine ants as democratic village states, while regarding the more advanced, expansionist myrmicine and formicine ants such as fire ants as totalitarian, with some elements of both communism and fascism. But Haskins saw the army ants as something of a special case, an early branch of the primitive ponerine species. Recent evolutionary assessments tend to show that they are more specialized, but Haskins's view of them, based on what was then known, presents a telling portrait. Haskins compared the army ants to early hunting communities that migrated over large areas to support themselves but did not conquer new lands. In any case, few large ant superorganisms appear to be influenced as much by their workers as the various species of army ants. More than other ants, army ants are societies in which the workers have played a much stronger role, not just in the choice of the colony queen and male but in the evolution of these individuals. Their size, their smell, their similarity to each other, and some of their habits have been shaped by the need to appeal to the workers.

No animal can choose its parents, but army ants are the closest to it. They can choose the parents of the next generation of workers who will carry on the genetic line of their superorganism. By the time of the next royal brood, normally a year away, most if not all of the current workers will be dead, but the next generation of males and virgin queens will be the progeny of the male and queen that these workers select.

This type of sexual selection—"arranged" pairings by the workers in which they get to choose both the father and the mother of their future nestmates—was first discussed by Nigel Franks and Bert Hölldobler in 1987. They proposed that the reason workers should be involved in choosing mates is that they will later invest in the progeny. With both males and queens, the large, vigorous individuals they choose probably tend to be the most robust and fertile. This may also partly explain the similarity of appearance in males and queens. Both seem to use most of the same outward characteristics to demonstrate their vigor and desirability to the workers. Both have a large battery of exocrine glands in their abdomens that produce perfumes to make themselves alluring. Since they choose largely on the basis of smell, the workers may have caused the perfumes to be similar in both sexes. And both males and queens have elongated, enlarged gasters, the males for their massive sperm vesicles and the females for their eggs. Their similarity in size and shape is therefore superficial; they are, after all, opposite sexes, equipped with completely different reproductive organs. And there is a profound difference in how long they live. The queen's body is built to live five or more years. The male lasts no more than a month. Males of many mammals as well as other animals usually die before females, but the army ant swarm raider queens live sixty times as long as the males. But do the workers, choosing their matriarch and patriarch for their matching characteristics, have a clue they are choosing a mating pair? To the workers, the crucial channels of communication—how the queen and male are perceived in smell and by touch—may be the same. Or, as some researchers suggest, they may well have acquired evolutionary adaptations that allow them to discriminate between the subtly different characteristics of males and females in order to optimize the genetic makeup of the offspring they will raise.

After the male and queen swarm raiders have mated, the male rolls off the female and he struggles to stand up. He feels shaky on his legs. His attraction to the workers is already fading. They no longer lick him or grab hold of him in their mandibles. He is still breathing. It may take an hour or a day, but the male cannot survive long. He has served his purpose in life. He lies there for twenty to twenty-five minutes before his world spins out of control. Rolling on his back, surrounded by the colony that adopted him and the queen with whom he has just mated, he quietly dies.

The workers wave their antennae over the dead male to investigate. The queen tries to nudge him with the tip of her mandibles. Then they all move away. Within a few hours the esters of death begin to be produced by his corpse. With this scent the queen and workers now receive the message of death. Six workers surround the bulky animal and pick him up. They carry the carcass off to the other side of the bivouac where there is a fresh pile of beetle cases and odd indigestible scraps from the migratory raid of the previous day. The carcass is deposited on this garbage heap, which is at the edge of the swamp land, flooded by the rains. But such a large piece of meat will never be left simply to rot. Running swiftly across the water is a male basilisk who has come into his own because the rains have more than doubled his foraging territory. Arriving swiftly at the trash heap, the lizard makes a quick swipe with his tongue and devours the object of his enthusiasm: one juicy male army ant. With the ant's exocrine glands still half-turned on, the lizard's mouthful has a strong taste. The lizard swallows the body quickly and darts away down the soggy trail and back to his riverside patrol. Later that night, just before midnight, the lizard hears the commotion of the swarm raiders moving the colony bivouac. He scampers up the bank to watch the dark procession, the view illuminated by a big moon rising over the river. The lizard even catches a glimpse of the queen herself—an oversized ant running on sturdy legs, workers all around her, licking and attending her, occasionally even jumping over her. Inside his belly the lizard carries the male ant, her mate, in the final stages of digestion. The spicy meal makes him belch again.

AS THE NIGHT PROCEEDS, things are stirring in the new leafcutter colony. The virgin leafcutter queen, now so eager to leave that she will chance the weather, has twice had to be restrained by the workers. She pulls them out of the colony, attached to her legs and wings, before they manage to drag her slowly back inside for a few more minutes. Her time is coming soon.

Chapter Seven

Nuptial Flight

She summons her wings for one final effort and . . . the ascending spiral of their intertwined flight whirls for one second in the hostile madness of love.

— *Maurice Maeterlinck*,

THE LIFE OF THE BEE, *1901*

The current reputation of the ant queen is derived from such old, abraded, toothless, timorous queens found in well-established colonies. But it is neither chivalrous nor scientific to dwell exclusively on the limitations of these decrepit beldames without calling to mind the charms and sacrifices of their younger days, for to bring up a family of even very small children without eating anything . . . is no trivial undertaking.

— *William Morton Wheeler*,

ANTS, *1910*

The hour for the nuptial flight draws near; a strange excitement pervades the ranks of the workers.

— *William Morton Wheeler*,

ANTS, *1910*

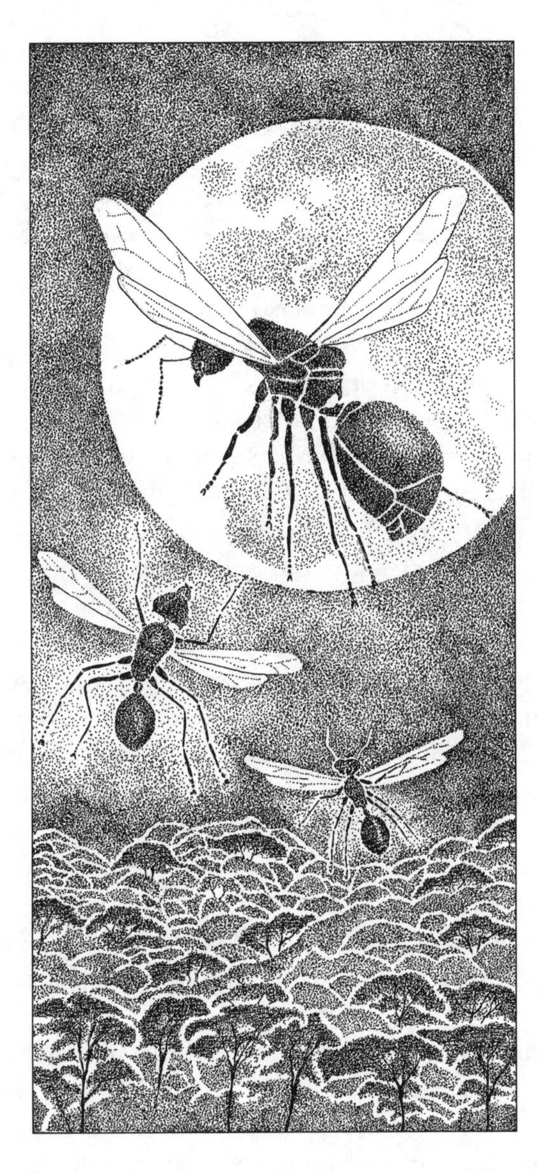

Nuptial Flight. Silhouetted by the moon, the virgin leafcutter queen flies high above La Selva. Two males approach and try to mate with her.

AFTER rain, there is life. Renewal. Rebirth. The memory of so much death in the flood quickly fades. The shuffled cards of life are dealt again. New players. A new game. And new things begin to happen. Almost all of the old canopy trees—vital anchors of the forest—have survived. And growing on them and all around them are fresh leaves, new epiphytes, green growth, flowers, and fruits. Nowhere are the changes more evident than in and around the light openings of the forest. A day in full sun makes things happen. In one typical opening at La Selva, *Heliconias* emerge in full bloom, attracting several hummingbirds. These hardy, striking plants send out various shoots up to twelve feet high from rhizomes just beneath the surface. At the end of each shoot is the "lobster claw"—red, orange, and/or yellow bracts from which protrude the greenish yellow flowers. *Heliconias,* which are in more and more demand as ornamentals by gardeners in warm climates, crave full sun, though the thirty Costa Rican species have adapted to varying degrees of access to the sun.

In another corner of this light gap opening are wild ginger plants, five to fourteen feet tall. Their spirals of long leaves are topped at the end of each stem by a yellow-and-red-striped flower that when closed resembles a pine cone. In fact, it functions like a pine cone, with the broadly overlapping bracts protecting the immature flower and fruit until the rainy season when the blooms appear, usually at dawn, and local traffic increases dramatically: Bees fly in to pollinate the flowers, and various ants crawl up to the base of the flowers to imbibe the extrafloral nectaries. Behind the ginger and towering above it is the tall, straight, grayish white laurel tree *Cordia alliodora.* It has shot up, seeming to have gained more than a few feet since flowering in the dry season. From the crown nearly eighty feet up, it pushes out hundreds of thousands of tiny, single-seeded fruits, parachute-fashion. And a special treat is the splitting open of the lime-sized fruits of the kapok tree, which had bloomed before the flood for the first time in ten years; the wind delicately blows away the brownish seeds loosely embedded in the downy kapok fiber.

This new rain forest world is suffused with the promise of new life. The luxuriant smells signal to the ants a return to their Garden of Eden. Some trees drop their leaves every year when their flowers bloom as a strategy to attract more pollinators who can then easily see the showy flowers. But many more trees keep their leaves or regenerate them constantly, ensuring the dense, lush greenness all around. Life in the rain forest revolves around alternating dry and rainy seasons. In fact, this break in the rain is not so much a dry spell as a return to the usual rainy season with its pattern of frequently overcast skies leading to a brief downpour every day.

But you wouldn't know it. The evening when clear skies arrive, the virgin leafcutter queen constantly pokes her head out of the colony nest. With her are sixty or so workers who have been restraining her as needed when her hormones almost drove her out prematurely, several times, no matter the heavy rain and floods. Now, however, the workers are excited, running in and out of the nest, pausing to lick the virgin queen, encouraging her to get ready. The message is that tonight may be the night. Surrounded by her entourage, she tastes the air and finds the humidity still high. The rain is gone, and she senses there is more light in the night sky than there has been for days. The moon, low on the horizon, is on the rise.

It is an ideal night for a nuptial flight. All around the forest several thousand other leafcutter colonies, those superorganisms that did not prematurely send out all their queens and males already, are making similar preparations. With any luck there will still be millions of ants meeting in dense swarms in the moonlit skies over La Selva.

About the time the queen is getting ready for the big night, her various male suitors, in colonies up to ten miles away, are also preparing for takeoff. From both the male and the female points of view, much is at stake. To the species, the males are more expendable—and worthless once they've mated—so this lends the individual male a sense of urgency to his behavior. The virgin queens have more opportunities, but they risk more in their repeated nuptial flights and their pairings with strange but attractive-smelling males.

Before the virgin queen leaves her colony home for good, she stops in the chamber nearest the entrance to grab some fungus. The minimas—the smallest ant workers—are busy harvesting a fresh crop. The virgin queen plucks a few of the kohlrabi pieces herself. In the darkness she chooses by touch and smell. It will be her last meal for months. After she eats a chunk of it, she balls up the rest in her mouth and tucks it safely into a tiny pocket under her tongue. It was the German-Brazilian researcher Hermann von Ihering who in 1898 discovered this secret of how the gardens of new colonies get started: The virgin queens carry the fungus with them. Those virgin queens who neglect to bring some starter fungus or who lose it in flight may mate, but they cannot start a successful colony.

As the virgin queen leaves the uppermost chamber, the queue of winged virgin queens waiting to depart extends down into the nest. She brushes past them. Accompanied by many attendants, she is escorted out of the nest for the last time. The same workers who were restraining her have released their hold and are now encouraging her to depart.

It is just after midnight. In the distance, the northern swarm raider army ant colony moves its bivouac on the first day of its migratory cycle. It scratches and hisses its way through the jungle night. It is a small but soon to be growing colony of swarm raiders, having just attracted its first guests—an ant bird and a few flies. But the leafcutter virgin queen has no concern about army ants now. When the moon begins to gleam in bright reflections on the leaves in the upper canopy, she jostles through the crowd of workers and soldiers stationed near the entrance of her colony and prepares herself to go. Some of the wet clay soil around the entrance sticks to her tarsal claws, and she cleans it off with her mandibles. The ground could be drier for takeoff, but the easy pliability of the earth will be useful when it comes to digging out her own nest—*if* she is able to mate and return to the earth to found a new colony.

Her chances are slim. According to estimates for a closely related leafcutter species found in the drier tropics, only one virgin queen in two thousand survives. That means 99.95 percent of the virgin queens produced die without founding a successful colony. She has already been tested in her response to her aborted virgin flight and to the flood. She has also had a measure of good fortune when the leafcutter scout intercepted the fly and saved her life. But the most difficult period of her life is upon her. Fortunately, she is well fed and has built up her food stores. Yet with so many queens and males having died in the heavy rains and in the flood, she now represents her colony's diminishing chances for getting its genes into the next generation. Still, she's a fighter with a fighting chance. Of course, there will be more chances with future broods of this superorganism in years to come, but the colony's immediate prospects for progeny depend on her and a few others. Following her out of the nest are at least fifty other assorted virgin queens and males, each with attending workers. They mill around the entrances and on the rooftop while the workers dance all around them, moving this way and that. It is a festive atmosphere—party time for the ants. Having dropped all their work, they wait for their winged brothers and sisters to take off. Some of them have never been out in the open air before. But it soon turns into an aggressive show for the young female workers. Several of them start acting like kamikaze fighters, ramming other insects who have gathered around to watch. Most are only curious at the spectacle. But the ant workers are hyped up, chemically stimulated, completely distracted from their work, and waiting for the air show to begin.

The virgin queen flutters her wings. She finds they have almost atrophied from lack of use for the past week. Will these things work again? She is going to need them for more than one reason. A predatory ground beetle, its mandibular lance drawn, has broken through the ring of workers and is advancing toward her. She again struggles with her wings, trying to move them up and down.

As if reinventing flight, she takes two steps forward and moves her muscles, contracting and relaxing them as fast as she can to set her wings in motion, vibrating, humming. She jumps up and takes off at a steep angle. She's airborne. In seconds the ground falls away, the predatory beetle snapping at her scent, which hangs in the air just above the ground. At the same time, all around her, her sister nestmates are also taking off, and the workers standing below look up, mandibles agape. Heading up moonlit shafts to the canopy, the virgin leafcutter queen climbs higher but then loses her rhythm, veers to the side, slips down, falls, a three-point landing. She's all right. Seventeen giddy seconds of flight, no more. She picks herself up, licks one dirty wing, and takes off again. This time she flies almost straight up like a rocket, as if impelled by the nearly full moon. It is a giant searchlight beaming over the Earth and a call to flight for a good percentage of the flying insects for hundreds of miles. Besides the moon beacon, all around her are the smells that rise up from the forest at night in the dank, steamy air, and the hums and whistles of thousands of stirring insects.

As the ground falls away, she sees below the males of her colony taking off and the workers already beginning to file back into the dark tunnels to return to their individual tasks. The excitement is over for them. Their energy will now be redirected toward colony growth for another year before they will once again focus on reproduction.

The virgin leafcutter queen flies a solo path, but her sisters are beside and behind her, never far. The separation increases as she flies higher. Soon no ant or other insect on the ground can see or smell her. Still, the stridulation, the sounds ants make as they fly—the same ones they make when buried under the earth—can be heard by many in the nighttime jungle. It is a high buzzing sound, which with a little imagination, could be the sounds associated with the early flight experiments of humans. Perhaps the males from other colonies, prowling the night skies above La Selva, are partly attracted by the peculiar noise, although most evidence indicates ants cannot pick up sound waves very well in air. As the virgin leafcutter queen flies along, she races past males and females, some of them joined together, entwined in mating and suddenly dropping out of the sky.

And then the virgin leafcutter queen notices him flying toward her. Out of the blue-black night he seems to have a bead on her. With his huge ebony eyes bulging out like black marbles set into his head, he has seen her first. He streams toward her, picking her out from all others. This male, along with several other males forming a small swarm, comes from a leafcutter nest in the southeast corner of La Selva. Since speed, good flying ability, and sharp vision are assets when it comes to finding and choosing the best mate in the night skies, these traits tend to be selected. He brings with him some 40 million to 80 million spermatozoa carried in the testes inside his abdomen. There is no evidence that males guard their sperm, but they should. Unlike many animals from fruit flies to vertebrates, ant males emerge from pupa into full maturity with all the sperm they will ever have. An ant male gets one shot and one shot only. Once he mates, there is nothing left. He is truly spent.

For ten seconds male and female fly a zigzag course, zooming and twisting through the night. The male starts closing in. He lurches now, ready to snatch her in flight, to possess her for a few eternal seconds. Their sturdy, reddish brown bodies dock. Chitin knocks against chitin, and he clings to her with his legs—yet they keep flying. She is full of his smell. He is growing drunk with hers. Then the male prepares to deflower his virgin queen. As his sperm moves from the testes to the vas deferens or staging organs, he grasps the female's thorax in his mandibles. From behind he presses against her gaster, massaging it with his forelegs. Fumbling around, he finally manages to insert his sexual organ at the tip of his gaster into the queen's cloaca, near the tip of her gaster. Almost instantly he gives her his sperm. It comes in a liquid base composed of secretions from his mucous gland. But he wants to make certain that he has successfully delivered his sperm. He does not immediately release his mandibular hold. With his legs, too, he clings fiercely. At that moment, through the flimsy, transparent wings of the desperate male, she glimpses other big-eyed suitor males approaching, three of them silhouetted in the moon. But she is already falling out of control. The entwined pair plummet several hundred feet —a spinning end-over-end freefall to the ground broken only by the queen's regaining some composure and stretching her wings to slow down at the last second or two before the crash.

For long minutes there is no movement. They are in a banana plantation some miles east of La Selva. Should the queen choose to found her colony here, she will have to contend with tent-making bats and perhaps in future years grazing cattle, if the rancher in the next field expands his holdings. She is a rain forest ant—much better suited to life in the jungle than to life in some cultivated field. Finally, after a few minutes that seem to last for hours, the male releases his grip. He is exhausted. He has done what he can to perpetuate his species, to get his genes into the next generation—his life's purpose. His life is complete. There will be no honeymoon. He limps over to a nearby banana leaf lying on the ground and slips beneath it. He is dying—even as she picks herself up to fly again.

The queen's night flight is just beginning. She lifts off. It's even easier this time, as if she has stolen energy from the spent male. She heads up and back toward the closed canopy forest, eager to meet more males.

NUPTIAL FLIGHTS ARE one way of getting ants from various distant locales to meet. With individuals mating outside their colony or superorganism, there is a high degree of outbreeding. This "mixing of the genes" gives genetic insurance to a population, improving its evolutionary chances to compete, survive, and flourish. Nuptial flights also provide a regular means of dispersion —much more common than being carried away in a flood or by stowing away in a ship or air cargo as a tramp species. More than one queen has been blown by brisk winds into the mountains to the cloud forests or over the high cordillera of the continental divide to the tropical dry forest along the Pacific coast—to distinctly other worlds. For a lowland rain forest ant, either of these fates usually means certain death due to desiccation in the inhospitable environment—inhospitable to them—or death by predators, whichever comes first. But if a few survive from time to time, perhaps at the margins of a different habitat, then the species' range will be extended and eventually new species may evolve in isolation.

Besides conventional nuptial flights in which flying males and females from different colonies meet in the air, there are other mating strategies. An hour or two before dawn on the same morning as the leafcutter queen embarks on her nuptial escapade, some two hundred miles away in Guanacaste province near Santa Rosa National Park on Costa Rica's Pacific side, a virgin queen bull's-horn acacia ant, *Pseudomyrmex ferrugineus,* departs from her parent colony in the shrublike acacia tree that has been her home. Taking wing, she flies a few hundred feet to the highest tree, the leguminous stinking toe tree *(Hymenaea courbaril)* at the top of a nearby hill. Perching there, she lifts her gaster in the air, pointing downwind. She releases a stream of sex pheromones and then waits. She releases more perfume, even more powerful, until the air downwind is saturated with the smell of an ant in heat.

The stinking toe is a good, quiet tree to perch in. Its mature leaves are avoided by various herbivorous insects and mammals. Even leafcutter ants won't touch them; an antifungal terpenoid in the leaves would poison the fungus gardens. This tree is also famous for producing most of the resin that fossilized the Neotropical amber. The acacia ant's genus, the fever ants *(Pseudomyrmex),* are ants often found trapped in amber. Some of them, no doubt, were calling at the time: Last call!

At regular intervals the acacia queen continues to release her pheromones, calling for males of her species who have already left their own parent acacias. The intensity of her smell says: There must be a few cruising around some distance downwind. Because it is dark, the swarms of males are pulled in largely by smell. Ideally, they need to find a female before dawn when social wasps fly out to prey on the big, clearly visible, helpless male ants, who can be easily plucked out of the early-morning sky.

Just before first light, several males, orienting to the intense perfume that seems to be coming from the top of the stinking toe tree, approach the female, sweeping upwind toward the scent. The quickest male with the sharpest sense of smell gets there first. He lands beside her on the branch, vibrates as he approaches, then seizes her and implants his seed. Instinctively, seconds later, the female jumps, pushing off the male in midair; they tumble to the ground.

He dies soon afterward. She flies a short distance, bites off her wings, and then sets off on foot to try to find an unoccupied acacia—particularly the bull's-horn acacias with the ready-made homes in the thorns and pearl bodies on the leaves that make them so appealing to ants. To find the right tree, she can search at least a month without food, but after three hours, she finds a seedling acacia nearby. There are already a few thorns with occupied founder queens. She climbs up, and in the first unoccupied two-pronged green thorn she finds, she cuts an entrance hole and hollows it out. Without waiting, she lays her first eggs. The race is on for her to build up her colony to be able to defend her thorn and eventually to expand. As the tree grows larger and other ants move in, the choice will soon become whether to launch raids against other queens in other thorns or be attacked herself. Eventually, there will be only one colony on this tree.

Unlike most other ant superorganisms that mate once a year, including leafcutters and army ants, the acacia ants can send out virgin queens and males throughout the year. Sometimes a queen will remain in calling position for several days before finding a male—subject to reasonable weather conditions and avoiding predators. The acacia ant makes up for the lack of conventional nuptial flights by being available any night of the year. The tradeoff is that she often gets stood up, but she's persistent.

Other female-calling ant species may be less successful than the fierce acacia ant. Standing on the ground at La Selva, the virgin queen of a locally rare ponerine ant, the primitive hunter *Amblyopone,* patiently calls for a male. Through the early morning no male even smells the call. *Amblyopone* is much more typical of female-calling ant species than the acacia ant.

Most female-calling ants are primitive ant species and/or social parasites. Typically, they have small colonies of twenty to one thousand workers and only a few reproductives. Each individual caller probably mates only once. In fact, this calling of a solitary female ant is reminiscent of the ant's wasplike ancestor. Even today many solitary female wasps call to their mates.

As female-calling ants, the acacia ants are unusual in not being primitive or socially parasitic or small. They support colonies, often of more than ten thousand workers, with two thousand males and a thousand virgin queens. But all the female-calling species are the same in their lack of reliance on synchronized departure and flights by reproductives of the same species or even the same colony. The winged virgin queens of the harvester ant of Florida—part of a group of ants that live underground, storing seeds—simply come out of their homes and walk around on the ground, broadcasting their pheromones. Males then fly in from distant areas to inseminate them. After sex, the females themselves *fly* off to start new colonies. Other queens that call are wingless. Sometimes the female must call with her pheromones for several days before a male finally comes to impregnate her. This may well have been the way ants early in their evolutionary history conducted their mating rituals. The many variations probably came later.

The most common ant strategy is the nuptial flight in which males get together, flying in swarms, and try to attract flying virgin queens. Depending on the species, the males typically come from many colonies, winging in to a particular site such as a forest clearing, a tree crown, a hilltop, or at the edge of the forest. Some species congregate in midair, flying slow circles or hovering a few feet above the ground. There is also some evidence that the same areas may be used from year to year. Ants using this mating strategy—referred to as "male-aggregation" species—tend to have large colonies that produce up to thousands of reproductive adults each year. The females are inseminated typically from one to eight times. But there are many variations on this mating method as well; for example, among the many *Pheidole* ant species, there are two in the southwestern United States: one that flies in the usual all-male swarms and one that flies in female swarms. In the male swarms, the females approach and fly in slow, even circles through the swarms of up to fifty males until a male zooms over to mount her in midair. They then plummet to the ground. In the unusual female swarms, the males are attracted by the pheromones for sex. Bert Hölldobler has watched the males fly into the female swarms to mate with individual females. And Costa Rican spider scientist William Eberhard has seen mixed swarms that range from mostly male to all female in the cattle-tending *Acropyga* ant, common at La Selva.

Still other variations are practiced by certain army ants such as the swarm raiders, who reproduce by fission—dividing the colony in two. The males alone fly, but it's only for dispersion. Once back on the ground, a male must make himself attractive and acceptable to a foreign colony and its virgin queen. As for little fire ants and a few other species, rather than nuptial flights or fission, they reproduce and found new colonies by "true budding," analagous to the way some plants reproduce. Each colony has numerous queens that are dispatched to found new colonies that retain their association with the parent colony.

With so many ant species—at least hundreds in a given tropical forest—how do males and females of the same species find one another and thus preserve their species' identity? Different mating patterns help, of course. The reproduction of all colonies of the same species is usually on a similar schedule. Most of the nuptial flights are tightly synchronized on a population or species basis—occurring after a certain time on a certain day and dependent on certain weather conditions. A clearing after a good rain at the start of the rainy season is one common choice as a time for reproductives to depart the nest; this is the strategy used by the big leafcutter ants. Such a signal—dampness in the air changing to dryness—is easy for an ant to read, and the softer ground after the rain makes it perfect for the newly mated queens to excavate their nests and start a colony.

But with so many species in one area, there are some overlaps. On any given night, dozens of ant superorganisms of different species might send reproductives out on nuptial forays. Also, with the female-calling species, the ants are often available for protracted periods and some of them even year-round. To make sure a male or female ant finds the right partner, the main attractant and means of verification has to be smell. The sex pheromones are unique to each species, which helps to ensure that when an ant displays its perfume, the right mates come to visit. Superficially, it resembles a singles bar where the various perfumes and aftershaves might attract different potential mates; of course, humans have visual and auditory clues as well to size up a mate. Ants must rely on smell. The subtleties of the attraction of each peculiar scent are such that a leafcutter ant is as different from an aztec ant as a spider monkey is from a human.

AFTER HER BRIEF visit to the banana plantation, the leafcutter queen manages to fly again to where a large number of males are swarming. In the next fifteen minutes, gaining altitude, she avoids dozens of odd suitors and predators who would try to mate with her or swallow her whole. One by one new males swoop in and dock with females that appeal to them. Then another exceptionally large male comes toward her, grabs hold, and they couple and fall. Following this pattern, she mates with the most persistent males, one after another, some from as far away as Nicaragua.

In all, the leafcutter queen mates seven times that night, storing the sperm in the special sac, or spermatheca, inside her body. After this one wild night, she will never mate again but will use the sperm she received from the seven males —some 250 million spermatozoa, give or take 50 million—to fertilize her eggs one by one, to make worker sisters year-round and potential new queens during her superorganism's annual reproductive phase. The sperm, kept alive as free-living cells in her spermatheca, will be parceled out as needed over her fifteen- to twenty-year or longer life—the life of this superorganism.

The obstacles that every queen must overcome in order to set up a successful colony are almost overwhelming. It is not surprising that few queens survive: On this night some five thousand leafcutter colonies dispatch more than 1 million virgin queens to fly the skies above La Selva and the surrounding region. Only a few queens might still be alive in a year's time. Some four hundred thousand might be eaten by various predators in flight such as bats, nighthawks, and other birds. Perhaps five hundred thousand others might manage to mate, spiral to earth with or without a male, but then be eaten by various predators or attacked and killed by ants from other colonies during routine hunting and scavenging.

Of the one hundred thousand left, thousands will certainly manage to mate, but in their impatience or failure to smell their mate or to recognize him by touch, they will couple with other ant species. If the species are closely related, the ants may be able to mate successfully, but their progeny will not survive. Yet matings with other species may be less disastrous if the queen mates more than once and does so at least once with her own species. This is the case of perhaps ten thousand of the leafcutter queens. Some ten thousand more queens will mate and fall into one of the area's rivers, all drowning except a handful that manage to crawl onto a large fleshy leaf that eventually lands ashore some miles downstream. At least ten thousand other mated queens will get caught in the canopy, landing in the bromeliads and tangled lianas and vines; those that are not eaten by the following late morning are overheated and desiccating, baked to death in the sun.

Perhaps five thousand other leafcutter queens will manage a successful flight, avoid every predator, mate six to eight times, succeed in digging out a nest and even laying eggs, only to find that they have forgotten or lost their fungus starter. In such cases the hapless leafcutter queen will often gather sand grains that are superficially like the fungus and proceed to fertilize them repeatedly with her own feces in an effort to get the fungus started. When the first brood arrives there is nothing to feed them. They can snack on the queen's food eggs, but eventually, with no fungus, they and subsequent broods will starve to death, and within a few months the queen will die, too. Researchers have found in laboratory tests that leafcutter queens without a fungus will sometimes accept a piece from another colony and use it effectively to get their own colony going, but this behavior may be rare in the wild. Since pieces of fungus from leafcutter colonies are not ordinarily carried outside the nest, the likelihood of a fungus-less leafcutter queen getting a second chance is slim to none.

The hazards are many, the chances of survival slim. In an effort to increase their chances of survival, certain gardening ants long ago adopted night flying for their nuptial endeavors. Survival might be even lower otherwise. The night fliers probably gain some advantage because there are fewer predators around.

The mating rituals of leafcutters stand as a notable test of fitness in the animal kingdom—a prime example of Darwin's survival of the fittest. From Wheeler to Weber to Wilson, all who have studied various aspects of the leafcutters' biology have been impressed. Leafcutters are superbly tested. And they are incredibly successful, dominating every ecosystem they inhabit.

Certainly, the once virgin leafcutter queen has been tested and so far has survived. This is the same queen who was saved by the scout from the phorid fly, the same queen who flew the abortive nuptial flight and then managed to return to the nest through the driving rain, the same queen who helped her superorganism evacuate during the flood and build a new nest and reestablish the colony on higher ground. Would she be among the few hundred queens who survive the night, never mind the five still alive a year from now?

There are still more challenges: With her last male mate, a male from a second growth forest beyond the banana farm five miles east of La Selva, she at last descends to the ground. Cushioning their fall is the thick leaf litter at the edge of a clearing in the northeast corner of La Selva—near the site of what will be a new 4,090-square-foot laboratory scheduled for completion in less than two years. Still entwined, both are exhausted; for a few seconds they lie in post-coitus. Then slowly, gently, the female pulls away without tearing off the male genitalia, as often happens with mating bees. The male never moves from the spot. He breathes his last as the female slips from his side and walks away.

She stumbles around for more than an hour in the predawn blackness beneath the canopy. She seems to be surveying the landscape, searching for something or maybe trying to get her bearings. She smells the trails of several smaller gardening ant colonies, but they are weak arms of some distant main trail. She notices that there are plenty of fresh green leaves following the rains, many of them growing on the fallen trees that have been pushed to the side of this clearing in the abandoned plantation site surrounding the developed part of La Selva where the bunkhouses and other buildings are. It is a somewhat disturbed site but is still good for leafcutters, with enough light to grow substantial amounts of vegetation. With all the rain, the leaf growth should be luxuriant over the next year—ideal for developing a strong leafcutter colony. There will be several million workers by the time the laboratory is constructed; they will have to deal with the increased human use of the area then, but at least they would be strong.

Minutes later she spots several bala, or bullet ant, workers scouting nearby and laying scent trails. Even though bala trails are mainly for individual ants, the last thing she wants is a trail that leads right past the entrance to her nascent colony. Today, of course, there are plenty of dead males to pick up. They grab one each in their mandibles and mark others to be retrieved later that morning. She is lucky. She stays frozen under some leaf litter and eludes the big balas' curiosity.

She must get started digging before sunrise. Now the swarm raiders and many others are silent, inactive. They and other animals and insects will emerge promptly at dawn looking for breakfast, and the bright sun, once it clears the upper trees and falls onto the clearing, will at best make her sluggish, at worst kill her. Already her wings are weakening at the base. Extending the wings at right angles to her body, the queen puts one of her legs behind each wing, one by one, to snap them off. One wing proves difficult. She alternates pulling with her legs and jaws, then rubbing against the ground to try to loosen it. Finally, extending her wing straight out from her body, she lifts one of her legs as far out on the forewing as possible and, with a quick jerk and an accompanying crunch, snaps the last wing free from her torso. Tossed on the ground, the wings are soon carried away by ant scavengers—the alert ants of various species that function as the rain forest clean-up crew. It is a marginal "find," however, because there is little nutritional value in the wings.

Now she is officially earthbound once more. The very action of dealation—the new queen losing her wings—imposes dramatic physiological as well as psychological changes, as Wheeler put it. "She is now an isolated being," he wrote in his classic 1910 *Ants* book, "henceforth restricted to a purely terrestrial existence, and has gone back to the ancestral level of the solitary female [wasp]." She will now rely on her hard-wired instinct—part of her genetic inheritance—for the elaborate plan she must follow to found a successful colony.

THE NEWLY MATED southern swarm raider queen, by contrast, has never had a period alone. Swarm raider queens rely on their ability to attract and keep great numbers of workers. After they become full-fledged queens with colonies, the failure to stay attractive and strong means abandonment by the colony and death soon afterward. It is only in the last days or hours of their lives that swarm raider queens are ever alone.

After another day's raid, the southern swarm raider queen's superorganism moves and settles for the night in a new bivouac. The queen is beginning the cycle that will lead in thirty-one to thirty-four days to her first worker brood. She is still active, but within days she will become seriously physogastric, her abdomen swollen with eggs. For the next few days she will launch the sort of massive daily raids that will support this new colony. Even with only half the worker force of a mature colony, and with the queen only a few weeks out of her cocoon, the new swarm raiders are able to produce and support a brood virtually the same size as the mother colony the first time around. Thus, because there are two new colonies, there are nearly twice as many brood only a month after the mother colony divides in two. The mature performance by young colonies is possible because of the intense grooming and feeding of the new queens.

Vigorous, highly charged, ravenous, the southern swarm raider queen's superorganism devours the jungle. On this first raid, some of the more delectable morsels they scavenge are the plump leafcutter males and unsuccessful leafcutter queens lying on the ground—as well as a couple of mated queens who failed to dig their holes and bury themselves in time. Besides the leafcutters, there are other ants to be picked up, along with the beetles and smaller arthropods, all of which are carried back along the trail to the booty cache. The swarm raiders are on a track to sweep across the ground where the new leafcutter queen has landed. She has perhaps an hour until the first fan of swarm raiders arrives.

FOR TOO MANY minutes in the early morning, the new leafcutter queen lingers, continuing to test the ground for her nest—pacing back and forth, at first in zigzags, then veering off erratically in a completely different direction. The queen ignores the rhythmic whoops of the oropendolas, the leaders of the dawn chorus of birds that wake half the jungle at the first sign of morning twilight and the start of the day shift. Somewhere over toward the blue Caribbean, where the little fire ants have survived another night on their raft, the sun is rising. As the sun looms into view through the trees, the daytime jungle lurches into full motion. In the distance is the rumbling of millions of footsteps—swarm raiders on dawn patrol. The queen will have to be careful. She needs to start digging. Still, the leafcutter queen wants just the right spot of ground—not too wet, not too dry—and won't settle until she has found it. If it is too dry, the queen will have to search for a damper spot or find water and bring it to her chosen spot, carrying droplets in her mandibles to water the earth until it has sufficient moisture to hold its shape when she digs out the tunnel and excavates the first chamber. If it is too damp, then she will be vulnerable to fungal attack, a pressing problem for lone ants that have no other insects to lick them, groom them, and keep them clean. Of course, she will do her best to keep herself clean. Alien fungi—unrelated to the fungus that leafcutters cultivate—are a problem for many insects in the moist lowland rain forest. Following all the hazards of the mating flight, almost 90 percent of newly impregnated leafcutter queens die in the few weeks in which they struggle alone to found a colony and start their first brood. Most of these deaths are due to a fungal invasion that they cannot protect themselves from—not from swarm raids.

She chooses a spot at the edge of the clearing near the base of a tree. In general, a leafcutter queen can avoid the routine digging of various mammalian predators by nesting near the base and among the roots of a tree. This is a young cecropia tree *(Cecropia obtusifolia),* one of the species commonly occupied by aztec ant colonies that fiercely patrol the tree and often cause havoc for other ants nesting even a few feet from the base. But it seems no one is home; there is no trace of the aztec odor. The fast-growing cecropia can also offer a ready source of fresh leaves. This is not why the leafcutter queen chooses this spot, but it could be a bonanza.

The earth here is soft from the rain, almost too soft, but better than too dry. Clearing the leaf litter around her chosen spot, she looks over to where she dropped her wings, only a few feet away. She digs straight down using her front legs and mandibles, excavating a tunnel up to half an inch in diameter and four or five inches deep. The soil is perfect! It is loose enough to work, yet has the right amount of dampness to hold the tunnel walls firmly in place. The queen brings the excess soil up to the surface in pellets, which she forms with her saliva and tosses out one by one; they pile up to form a neat crater wall around the entrance, like a miniature volcano. It takes half a day of hard labor to dig out the main entrance and tunnel to where the first chamber will be.

The leafcutter queen is down her hole when one small, bedraggled aztec superorganism files past and heads up the tree trunk—eight workers in all, two carrying homopteran cattle, six flanking the queen on all sides. With great skill the few surviving workers take turns gnawing at the pit below the first leaf base to gain entrance to the chamber inside the trunk.

Later that evening another aztec queen crawls up the tree, passing the already occupied chamber and finding her own spot a few feet up. Recently returned from her nuptial flight and without a workforce, she does the gnawing herself, perforating the trunk within an hour. Once the hole is wide enough, she moves in. Then comes the remodeling as she fashions a spongy nest from the brown waxlike carton material used by many aztec ant species, whether they live inside living trees or outside in homemade nests attached to branches. The carton is made from plant fibers and secretions from the workers, or in this case from the queen herself. The aztecs that make them outside the nest typically affix them to the trunks of large trees. In size and overall appearance they are not unlike an eagle's or large bird of prey's nest. They can be up to three feet across and more than seven feet long. These large external carton nests are thought to be simply an extreme extension of the common use of carton for nest modification inside trees such as cecropia. Within days she will lay eggs to start her own aztec colony.

The Ant and the Anteater. As the leafcutter queen excavates her nest beneath the cecropia tree and starts her fungus garden, a male tamandua, or banded anteater, attacks.

Meantime, back on the ground, the first investigations of the leafcutter's activity begin. Few enterprises in the rain forest, even the intentionally surreptitious founding of ant colonies, manage to escape all notice from the neighbors.

The aztecs, however, are busy founding their colony and have not entered the advanced chest-beating stage when they might patrol the base and extend their empire to the area surrounding the tree. Certainly, the leafcutter queen knows nothing about the aztec queen, though the pungent odor of cyclic ketones that aztecs use for their alarm pheromones hangs in the air, and it's only a matter of a few weeks before the first leafcutter foragers find out. No, the intruders are not the aztecs. Ground-living creatures on an early-morning prowl are the first to run across the leafcutter queen. She spits a dirt pellet out the entrance hole and senses the flashing interruption in the shafts of light streaming down through the open canopy. All of a sudden, poking through the entrance above, a giant snout sends an overpowering gust of wind down the tunnel. She is pinned against the wall. Then a massive sticky-wet tongue comes charging down. It misses the queen by inches and collects dirt instead. A hungry anteater. This is a male tamandua, or banded anteater *(Tamandua mexicana)*, a small anteater, only twelve pounds when mature, with golden brown fur and a black V across the back that looks like a vest worn backward, but to an ant it is truly a giant monster. Besides the claws and the long saliva-covered tongue, anteaters have specially adapted, thick-walled, muscular stomachs for crushing the hard exoskeletons of ants and other insects. The chemical defenses of most ants do not bother anteaters much. In a pinch they'll even go after aztec ants up cecropia trees—until they are driven away, rubbing and scratching themselves like crazy. The only ants they seem to avoid are the larger ponerines and army ants. In all, about 40 percent of their diet is ants. In fact, certain anteaters, such as tamandua, tend to eat more termites than ants; it depends on the individual animal's preferences and foraging patterns whether they usually hunt on the ground or up in trees. Active by day and by night, tamanduas hunt by scent, which is a good way to go when you're looking for ants and termites in the soil and rotting wood. When it comes to leafcutter ants, tamandua find them spiny and less than palatable. But the day after a nuptial flight, a new queen full of eggs and sperm is a substantial hors d'oeuvre. Various native peoples have used the leafcutter queens as food, as did early nineteenth-century settlers to South America. Leafcutter queens, who must subsist on their own fat bodies for months, are renowned for their plumpness even compared to other founding ant queens.

The leafcutter queen crawls to the deepest part, the base of her tunnel, and stays motionless. She is deep enough to avoid being eaten, but will the big mammal decide to try to dig her out with his powerful foreclaws? Maybe she should have chosen to dig her nest closer to the base of the cecropia tree, but then she would certainly have risked future dangerous encounters with the aztecs.

The leafcutter queen claws sideways, burrowing into a narrow crevice and covering herself partly with dirt. In a mature leafcutter colony, soldiers would be dispatched instantly to mount a fierce attack on any invader, whether an anteater or another ant. After a few hundred nasty bites, if not before, the anteater would usually find a less aggressive ant colony to bother. But the lone queen, her first worker brood yet to be laid and more than a month away from maturity and many months away from producing her first soldiers, has few options but to sit tight. The fungus to be cultured is still inside her mouth—if she hasn't lost it.

The anteater starts digging but promptly hits a root. Again he tries to slip his tongue down the tunnel to probe for the juicy ant prey, but the queen remains absolutely still. Then, as suddenly as the giant animal arrived, it disappears. The leafcutter queen hears the heavy footfalls of the southern swarm raiders who chase off the anteater. For once the thundering swarm raiders prove useful. She quickly seals off the nest. It is dark in her tiny bunker, but she is safe. The swarm raid seems to last for hours.

To the young leafcutter queen, a swarm raid over her colony is mostly good news—once she has survived. They won't touch the greenery and actually remove many plant-eating insects that compete with leafcutters—though none of them present much of a challenge. Also, a swarm raid means that many of the potential predator arthropods are temporarily displaced or put out of action. The leafcutter superorganism can go about its business without having to worry much about defense—even though more ants will soon move in, starting with the more aggressive, opportunistic species. Finally, a swarm raid means, quite simply, that the swarm raiders themselves will not come calling again for some time—months or even years.

Compared to the leafcutter queen and her risky attempts to start a superorganism from scratch, the new swarm raider colony is only a few days older but is much stronger, larger, and more active. The army ant strategy of splitting in two seems less risky, though of course fewer new colonies can be produced this way. If the leafcutter superorganism—which dispatches thousands of reproductives into the world—is successful in producing more than one new colony a year, the mother leafcutter colony then surpasses the usual reproduction rate of a successful army ant colony.

Yet in some ways, like the leafcutter queen, the southern swarm raider superorganism is still taking its first steps in the world. The new superorganism has several hundred thousand workers borrowed from its mother, but where are the notorious camp followers—the many other species that travel and live as honored guests of the colony? These myrmecophiles, or ant lovers, live in varying degrees of symbiosis with the swarm raiders as part of the unique swarm raider army ant ecosystem. The army ant–loving symbionts vary, from species that do something for the swarm raiders on an exchange basis to outright parasites. When the mother superorganism divided in two, however, many of the guests became dispersed and were lost or wandered off, joining other superorganisms. It will take months to build up each swarm raider colony to its formidable original size and perhaps even longer to build up each swarm raider ecosystem.

Setting out on the third day's migratory raid as a new colony, the southern swarm raiders run into hundreds of mites along the trail as well as numerous flies and several attractive rove, or staphylinid, beetles. Attracted by the plentiful food, all are eager to join the swarm raider superorganism. The swarm raiders hardly notice. Keeping a steady pace, the workers are occupied in the excitement of the morning's raid. Many of the workers are killing other insects they encounter, biting them with their sharp jaws. But they are also scavenging, snapping up scraps of meat: legs of various arthropods, termite heads, and numerous whole ants of various sizes and species. At the same time, other workers are forging ahead. It is turning into a good day's raid.

As the army ants drive forward in their characteristic zigzag fashion, they barely feel the odd tiny mite in the leaf litter grabbing hold of a leg here, an antenna there, climbing aboard and gravitating toward favored spots on the ants' anatomy. Mites are small even to ants, and individuals of many species can fit comfortably on the head of a pin. Others may even be as large or larger than a pin head. There are as many species of mites as there are conceivable live-in situations for mites. Of course, many workers are already carrying mites. For many mites and their ant hosts, it is a permanent arrangement, "till death do us part." The most specialized mites probably never have a free-living existence; they are born, reproduce, and die without ever leaving the bodies or nests of their hosts. But they may move around within that world.

The various species of mites crawl to their favored locales and settle in. There is one mite that prefers riding on an ant's head, another on its thorax, still another on the gaster. Many of the mites that live with army ants have special modified claws or teeth or other ridges to enable them to hold on to particular parts of the swarm raider anatomy. One mite prefers to ride on the inside of the mandibles where it nestles safely behind the big ice tongs of the large workers and soldiers. Still another, which specializes in column-raiding army ants, attaches to an ant's foot and even performs the functions of the foot for the ant. When army ants form bivouacs, bridges, and clusters by hooking their tarsal claws to the leg of another ant to join together, the workers with one or more mites on their feet often form strands using the hind legs of the mite in place of their own tarsal claws. Because the mites are small, the entire hind legs of the mite must assume a curved position to form the claws.

Mites live on ant workers, adult males, queens, and even larvae. Many ant workers carry several mites. Some mites feed on blood as parasites while others may obtain part of their nourishment by licking the oily secretions on the ant's body. Still others feed as scavengers within the army ant nests. These mites, when found on the ants, may just be hitchhikers, jumping aboard for a ride back to the bivouac. Their ultimate destination is the refuse piles or kitchen middens of the ants, where they will embark on a rich life subsisting on the substantial scraps of waste created by life in a rapidly expanding army ant colony.

As the southern swarm raiders reach the full extent of their raid, a young male beetle, a rove beetle about the same size as a swarm raider worker, waltzes in from the edge of the leaf litter and boldly grabs a waiting young worker, standing on the fringes, from behind. The beetle grabs the base of the ant's antennae in his mandibles, rendering her somewhat helpless to shake off the intruder. The beetle quickly maneuvers into place, straddling the ant from above and slightly to the side, his long torso over the ant's back and his legs locking around the ant's slightly larger abdomen. The ant becomes a bucking bronco. Riding sidesaddle, the beetle proceeds to rub the ant all around her body, top to bottom, from head to gaster, with his legs—an attempt to pacify the ant. The rubbing proceeds slowly at first, but the reaction is almost instant. The ant calms down and soon stands limp, accepting whatever the beetle has to offer. With six legs alternately rubbing, the massage covers the ant's entire torso and head many times over. Gradually, the strokes begin to alternate between the ant and the beetle's own body, as the beetle changes his tactic. The beetle effectively transfers surface oils and scents on the ant's body to the slightly sticky pads on his feet and from there to his own body.

The beetle's rubdown of the ant provides a clue to the way some insects are able to become full-fledged guests of army ant colonies. Army ant specialist Carl Rettenmeyer from the University of Connecticut suggests that using this grooming method, the guests of army ants are able to pick up the colony odor, transferring it to themselves as they stroke first the ant, then themselves. In time the beetle acquires the ant's individual and colony odor. It may take hours or even several days. Other beetles, however, may be able to earn acceptance by simply curling up in their shells inside the ranks of army ants, withstanding predation, until they have acquired the right scent. Still other beetles, not necessarily associated with army ants, offer special appeasement substances through a gland at the rear of the abdomen. The ant licks the gland presented by the beetle and, once appeased, moves along the beetle's abdomen to the "adoption gland." After licking this gland, the ant carries the beetle into her nest. The transfer of chemical substances—whether through grooming or direct communication—is a key step in a guest's bid to obtain a passport into a swarm raider ant colony and to become accepted.

The mites and beetles—loosely called army ant "camp followers"—are just a few of the many arthropods that are myrmecophiles. Among all ants there are certainly thousands of such species. In general, the bigger the colony, the more guests. With swarm-raiding army ants, who have some of the biggest colonies, it is a combination of sheer size as well as flamboyant predatory behavior that attracts more guests than any other ants. There is even a special name for army ant guests—a subclassification of myrmecophiles called ecitophiles—literally, lovers of army ants. Hölldobler and Wilson list twenty-five families of arthropods that ecitophiles belong to. These includes various families of mites and beetles, flies, silverfish, whip scorpions, and butterflies. Each family has one or more species that sometimes or often become ecitophiles.

ANT GUESTS ARE a big topic in entomology. There have been substantial papers and books, even whole careers, devoted to the guests of ants, though much of the extensive literature is buried in papers on taxonomy, behavior, and ecology of ants and other social insects. Father Erich Wasmann began the modern study of ant guests in 1894 and returned to it often during his long career. Wasmann divided myrmecophiles into five behavioral categories that are still used today to show degrees of integration into the ants' social system:

- The first group are arthropods that ants dislike and treat in a hostile fashion but that thrive as predators along with the ants due to their superior agility or defensive ability.

- The second are also predatory arthropods, but these are ignored by their hosts primarily due to their neutral odor.

- The third group—sometimes called "true" guests—are symbionts accepted at least to some extent by the ants as members of the colony.

- The fourth group, the parasites, includes those arthropods that live on the body surfaces of the ants and feed on their excretions as well as those that penetrate the exoskeleton to feed either by tapping underneath the hard shell or by moving inside entirely.

- The fifth group, called trophobionts, include the insect cattle and caterpillars that live on plants. These are the arthropods willing to supply ants with honeydew or other secretions in exchange for protection from parasites and predators and relief from their honeydew loads.

Yet these categories have proved not as clear-cut as originally conceived. Some myrmecophiles not only live with and feed their hosts but sometimes eat them or their brood. Symbiosis—which in its modern scientific definition includes any intimate, protracted dependent relationship, no matter who benefits—can be a complicated business.

Entry to an ant superorganism, much less acceptance, is tightly regulated—largely through chemical signals, though touch probably plays a minor role. To work their way into the colony, the eager myrmecophile may embark on "elaborate grooming rituals" or chemical tomfoolery. Using special olfactory passwords, members can enter an ant society and travel with it. Through evolution, certain insects have been able to break down the barriers, some even managing to set up permanent symbiotic relationships with ant colonies in which the ants treat them as sister nestmates. In Hölldobler and Wilson's words, these myrmecophiles "have attained the supposed ability to speak the ants' language of mechanical and chemical cues." They have broken the code.

While working on museum specimens of ants and their guests, Wasmann first observed the phenomenon that some beetles look remarkably like various army ants in body shape and color. The common beetle profile tends to be round or chunky in the middle, but the army ant guest beetles have narrow waists that resemble the characteristic petioles of ants. The beetles' entire bodies can be slender like the ants'. Their color even matches. Wasmann believed that the beetles were able to deceive the ants into accepting them into the colony. He also pointed out that the coloring might help deceive antbirds: The beetles would look like the ants and thus avoid being eaten. But more than that, he pushed the idea of "tactile mimicry." Tactile mimicry, which later came to be known as "Wasmannian mimicry," refers to the ability of the beetles to deceive the ants by closely resembling them to the extent that ants, running their antennae over the beetle torsos, decide that they are ants and admit them to the colony.

In the battalions of parasitic beetles that march with army ants there are indeed many species that look like their hosts. Yet whom is the mimicry intended for, or, as Wilson says, whom does the mimicry fool? David Kistner from California State University presents strong evidence for some tactical mimicry in certain ant-beetle associations, yet Wilson, Hölldobler, and others favor the idea that the beetles' visual mimicry is aimed only at the antbirds and other predators that follow swarm raider columns. The closer a beetle resembles an ant, the more likely it escapes being eaten and is able to reproduce. Also, mini army ants that camp underground have no beetle mimics; only army ants such as the swarm raiders that raid on the surface where they are visible to antbirds have these beetles among them.

Rather than using tactile mimicry, it is now accepted that beetles gain entrance to an ant colony through chemical passports. Hölldobler has experimentally changed the smell of a worker ant in a colony. It doesn't matter what she looks like, she is banished or killed. The key is whether the myrmecophile guest smells like an ant and can "talk" like an ant. In other words, with the beetles and probably most if not all guests, the ant colony must be fooled into thinking that these foreign insects *are* ants.

Wasmannian mimicry has recently been extended to cover "all mimicry of social releasers," according to Wilson and Hölldobler, "including behavioral imitation of solicitation signals and the imitation of ant pheromones by myrmecophiles." This is at odds with Wasmann's original meaning; it contradicts his theory about amical selection and symphilic instincts. Symphilic instincts refer to the tendency to adopt and nurture organisms other than the ants' own species, a tendency derived from the instincts involved with nurturing of the brood. It is the ants' symphilic instincts, according to Wasmann, that led to amical selection, a kind of artificial selection exerted by ants on their symphiles with the result that the symphiles become more and more dependent on their hosts and more specialized for the role. These Wasmannian notions are no longer pursued or even entertained. Yet Wilson, Hölldobler, and others agree that the expanded concept of Wasmannian mimicry is reasonable: It honors the name of a great myrmecologist who was the most important pioneer in discovering the world of symbiosis in insect societies.

Besides Father Wasmann and Theodore Schneirla, several other swarm raider authorities have spent time watching and researching army ant guests—as much or more time than they have spent studying army ants themselves. Carl Rettenmeyer, working at first with Schneirla and then on his own, devoted his Ph.D. thesis to the behavior of the various arthropods associated with New World army ants. He later wrote separate papers on flies, mites, millipedes, and silverfish. His work from Panama to Kansas provides some of the most detailed glimpses we have into life among the army ants and their guests.

But even a few years before Rettenmeyer, a Franciscan priest living in Brazil, Thomaz Borgmeier, did taxonomic work on army ants, as well as leafcutters, fire ants, and others throughout Latin America, writing mainly in German and Portuguese. He has his name on more Neotropical army ant species than anyone else. Through his army ant studies he became deeply interested in the various guests of army ants, especially the phorid flies of Latin America. His story has many of the bizarre but fascinating twists and turns of a twentieth-century myrmecologist. In the late 1940s his career was almost cut short by cancer. Neotropical myrmecology—for some reason blessed with a surprising number of priests, including Wasmann and Walter W. Kempf—sought divine assistance. Borgmeier decided to make a *promesa,* a sacred promise, to the Virgin Mary that if he were allowed to live and carry on his work, he would give up the editorship of his beloved *Revista de Entomológica* (The Magazine of Entomology). Under his stewardship through the 1930s and 1940s, this journal about ants and other insects was respected and achieved a goodly number of subscribers. It was clearly Borgmeier's pride and joy. After an operation to remove his penis, where the cancer was located, he was spared. He dutifully surrendered the editorship of his journal and returned to his ecclesiastical duties. He was put in charge of a home for blind women up in the hills near Rio de Janeiro. Concentrating on his ant studies, he published a few short ant papers in a Brazilian biology journal. At the same time he was working on a major extended work on army ants, and he had no idea how he was going to get it published. He wanted to make sure it was done right. He discussed the matter with W. L. Brown, with whom he exchanged visits several times alternately in São Paulo and Ithaca, New York. Brown remembers a larger-than-life, "charmingly disingenuous," yet dedicated scientist. Borgmeier had once been written up and photographed in *Life* magazine, which made him a celebrity among "the brothers," as he called them. Finally, the solution came to Borgmeier as if in a vision: He would start a *new* journal. And so up in the hills around Rio, living in a convent, with regular trips to the countryside to study ants in his late-model, Brazilian-made Volkswagen Beetle, Borgmeier finished his 716-page opus on the *wanderameisen*—the army ants—of the Neotropics. He published it in the third issue of his new journal, *Studia Entomológica* (Entomology Research), in 1955. Eight years later, in the same journal, he published his 367-page catalogue of the phorid flies of the world. His was the first detailed account of these most numerous guests of the army ants —an astonishing guest list that extended the known myrmecophiles, especially ecitophiles, to many more species.

AS THE SOUTHERN swarm raiders file back to the nest, they are no longer interested in capturing prey. They are moving slower now; they are subdued. It's an opportunity for the phorid flies to sweep in and begin dive-bombing. Most of the raiders have their "hands full," having picked up meat from the booty caches established along the trail to transport it back to the bivouac where the queen and most of the soldiers wait. But despite the army ants' seeming vulnerability, the flies are not trying to lay eggs on them as they do with many other ants, such as leafcutters. Instead, the flies target a few straggler non-army ants and other species who somehow eluded capture in the morning's swarm raid. A few of these are beetles who run outside the main part of the swarm raid but—minus the Wasmannian mimicry—stand out clearly from the ants and make for easy pickings. As these unfortunates scatter before the returning swarm raiders, the flies dart off after them, landing on their backs, and depositing eggs or larvae on them. The flies don't touch the prey captured by the army ants or the army ants themselves. That's the deal. In time, as the superorganism grows to maturity, the flies will build up to such numbers in association with the army ants that small mammals or lizards caught by the advance will look up to see the skies above the swarms almost dark with flies and hear buzzing that blocks out all other sound. In the swarm of flies is one species, *Borgmeieriphora kempfi.* The name, lost on the fly, is part of the legacy of a lifetime researching these flies.

Nearing the bivouac, the smaller workers mill about outside as the swarm raid dissolves. As the workers' bodies form into the bivouac, which begins to hang in strands from a low branch to form the walls, the ants metamorphose into a pleasing tapestry. The smaller workers' yellow to orange gasters contrast with the reddish black colors of the thorax—which is the same color of the larger workers and soldiers. Now the full ecosystem becomes evident as the beetles, mites, and flies sort themselves out and gravitate to their chosen spots on the animals or in the middens. One way to see the swarm raider colony is as a superorganism—a giant single animal on the prowl; the other is as an ecosystem, a partially isolated ecological island, complete with thousands of insects and other arthropods. Either way, the aspect that clearly marks it is its success. Swarm raider superorganisms, as ecosystems, attract a wide variety of animal life looking for an easy ride or a role in life. Each is looking for its own private microhabitat, its own niche, in the swarm raider ecosystem. The microhabitats of swarm raiders are rich and diverse, and according to Hölldobler and Wilson, include foraging trunk routes, refuse areas, peripheral nest chambers and guard nests, storage chambers, brood chambers with separate areas for pupae, larvae, and eggs, queen chambers, and the bodies of adult and immature inhabitants of the nest. All army ant superorganisms—because of these diverse microhabitats, their numbers, or the sheer density of their colonies, as well as their longevity—attract a lot of myrmecophiles, probably more than any other ants. Many become specially adapted to their niches. They can orient to army ant odor trails—with or without visual or tactile cues from the army ants! Swarm raiders because of the large size of the colonies and their migratory habits attract by far the most myrmecophiles of all the army ants, and probably of all ants. As the superorganism grows larger in size month by month, a new brood of two hundred thousand eggs laid every thirty-five to forty days, 2 million eggs laid a year, its guest list, too, grows.

The following morning, as the superorganism departs for another big raid, the beetles and flies are with them, becoming full-fledged members. The guest list now stands at about three thousand individuals of various species. Yet even more illustrious visitors are about to join up: the antbirds.

First a spotted antbird—a dark bird about the size of a house swallow—flutters out of the brush and swoops low over the flanks of the advancing swarm raiders, snatching up insects flushed out and now on the run. Then comes another species, the bicolored antbird, working the edges of the swarm just like the first. Finally, a third "antbird," a black-faced ant thrush, struts in, its short tail cocked upward and its head held high. It is whistling a simple downscale *shee-yoo, shee-yoo, shee-yoo, shee-yoo.* These three birds are among some twenty-eight species of "professional antbirds" in the Neotropics, all of which feed on the arthropods flushed out and scattered by army ants. The three antbirds often found at La Selva are typical and common companions whose very association invariably means that army ants are near. The only time they may stop following the ants is during the breeding season. Then each pair, which generally mate for life, cooperates in building a nest. However, if army ants enter the birds' territory, the antbirds quickly join up for a morning raid. They rarely if ever eat the army ants. The birds also miss the guest beetles that mimic the army ants. But the beetles and other arthropods that stand out from the swarm—with no special adaptations to look like army ants—make tasty provisions for the antbirds. Each species of antbird has its place or specialty in the advancing swarm—center, edges, on brush, or on the ground. Antbirds appear to be a drab lot compared to other tropical birds. Yet Alexander Skutch, Costa Rican naturalist and birdman, found a great diversity of tropical antbirds with subtle differences in foraging, plumage, and calls. And there are even more antbird species in the Amazon than in Costa Rica.

And as if to put on the finishing touches, by late morning half a dozen female army ant butterflies have boarded the ecosystem. These are one of the largest tiger-stripe butterflies. Their yellow, black, and orange coloration serves as a warning for birds that they are unpalatable—due to the poisonous alkaloids that they incorporate as larvae from their host plant. Because birds avoid them, the tiger stripes are widely imitated as a color pattern in various kinds of butterflies and even moths. Even though these mimics are edible, they look poisonous. Army ant butterflies frequently get together in mixed-sex courtship groups of up to a hundred individuals, but the business of following army ants is done only by the females. This association was pointed out as recently as 1976, and there are various theories about it, mostly having to do with the butterflies being attracted to the swarm raiders' odors. The butterflies feed on plants but apparently supplement their diet by licking the whitish droppings of the antbirds. As Bill Brown says, there is usually at least one animal ready to come around and clean up another's mess—something that will eat anything—especially in the tropics. Lapping up the bird droppings, the butterflies obtain valuable nitrogen compounds, which the females can use for egg production.

Now at last the southern swarm raider superorganism, the new daughter colony, is beginning to look like a real swarm raider colony. It is still comparatively small, but all the players have arrived. A new dominant colony—a new ecosystem—has truly been founded in the lush lowland jungles of La Selva. Over the next couple of years, the number of workers and guests will continue to expand. The superorganism will soon flex its muscles to have an impact on almost every square inch of its world.

FAR FROM COSTA RICA, colony founding could not be more different for the little fire ants—compared to army, leafcutter, and aztec ants. Conquering ants can afford to be patient. In their protracted attempt to set up a new colony, they are still waiting to make a landfall. After the Corn Islands of Nicaragua they were blown east for several days. Some 150 miles east of Nicaragua and 250 miles north of Costa Rica, they wash up on the beach on Isla de Providencia, a seven-square-mile inhabited island owned by Colombia, coming to rest on a small beach in the midst of the island's mangrove forests. The absence of motion is a novelty, and no ant moves through the night. Most of the vegetation is second-growth forest and mainly the same plants that grow on the Costa Rican Atlantic lowlands. Dan Janzen visited here in 1969 and found a paucity of insects, both in terms of numbers and species. But he did find one of his favorite acacia ants occupying the same bull's-horn acacia tree species as on the mainland. He hypothesizes that the seeds of the acacia tree arrived on the island first, carried by birds. Later, queens from ant colonies found in Panama, Nicaragua, or Costa Rica flew in or were blown in. Cecropias are also found here, but without the aztec ants these cecropias have lost the ability to provide food bodies. They and the acacias with no resident ants are doing fine here, however; there simply aren't that many insects. It would seem to leave things wide open and provide a perfect opportunity for the little fire ants to move in. Their only caveat is to avoid ant-acacia trees. But they never get to step off the raft. In the morning the highest tide of the month starts rocking and rolling the little ship, and they slip back through the light surf on their long, eventful odyssey.

On day seven at sea, the superorganism drifts north off the Gorda Bank northeast of Honduras and enters decidedly deeper waters. Crossing the Cayman Trench, two miles or deeper in spots, the makeshift schooner is holding up well and is now on a course to miss Grand Cayman altogether.

The weather, too, is holding, with full sun, and the sea has only a light chop. That afternoon the little fire ant colony encounters one of the marine water striders, or sea skaters. Most of these live close to shore, but there are a few, closely related to the semiaquatic water striders, or pond skaters, that have colonized the open ocean and have adapted to life even far out to sea. Dark with silvery markings and small abdomens with no wings, they move about on the surface of the sea, feeding on surface animals such as jellyfish. They lay their eggs on floating objects, including driftwood, feathers, or even birds resting on the waves.

With her large, globular eyes, the sea skater studies the fire ant raft. The ants are down below, invisible. Crawling aboard, the sea skater starts to lay her eggs when a fire ant appears and sounds the alarm. The eggs are snapped up like CARE packages dropped from the sky. The little fire ants love to scavenge, but the past week has offered slim pickings. Now there's a feast laid out on deck, and the centerpiece is an insect they've never encountered before. One ant manages to get her mandibles around the slender spiderlike foreleg, but the sea skater merely slips off the raft and shakes the ant free in the process. Another watery death. There are now only sixteen left; still, the queen is physogastric, and all the colony needs to survive is the queen and a handful of workers.

THE YOUNG LEAFCUTTER queen, settled beside the small cecropia tree near the site where the new lab will be built, readies her primitive self-constructed brood chambers. In the darkness of the sealed-off nest she opens her mouth and tries to expel the fungus that she has carried so faithfully ever since she left her colony. It's there! She maneuvers the wad from beneath her tongue and spits it out on the ground in the center of the main chamber. Then she bends her abdomen beneath her legs and touches the wad with the tip of it, squirting out a brownish yellow drop of excrement. Fertilized, the fungus begins to grow. Shortly afterward, she lays her first egg in the corner of the chamber, away from the fungus for now. She lays another and another—three to six in the first batch. For the next two weeks, at intervals of an hour or two, she will nourish the fungus garden with her fecal drops, tearing off bits and touching them to her abdomen, then replacing the bits. Meantime, she lays more eggs. Most of these she uses for her own food or saves to feed the workers. She keeps perhaps twenty or thirty fertilized eggs aside to be her first worker brood, but she waits until her garden is flourishing to bring the garden together with these eggs that will become her first workers.

After only a few days, there are problems with the garden. The fungus is growing poorly, and more fertilizer is urgently needed. The queen seems to know instinctively what to do. It is not always needed, but this time it's a matter of life and death for the colony. Climbing up the dark main tunnel chamber, the queen knocks the plug of dirt away and slips out into the daylight. She pauses to look around. She waves her antennae from side to side to catch whatever smells come wafting across the forest floor. She starts to search for a quick source of leaves for the garden. The first ones she encounters are from an unpalatable tree. She is repelled by the smell and knows these are unsuitable. Leafcutter ants commonly avoid collecting leaves of various species loaded with defense compounds like the stinking toe tree on the Pacific side of Costa Rica, but there are many others. A safe bet is often a younger tree with tender leaves.

She walks on. A few feet farther, near the path the researchers sometimes take, she finds a young tree. Smelling and then tasting a bit of the leaf sap, she confirms that this is one of the preferred leafcutter species. Now that she has gone out and found the food, as the scout used to do, she must use her jaws to cut some leaf pieces and bring them back to her nest. She must rely on her ancient origins from a time before leafcutters developed their intricate assembly lines and divided the labor of the colony so adeptly. A large physogastric ant, she must nevertheless manage to scamper up the branch, cut out the leaf, and carry it back to the nest before any number of predators come along.

She slices off a piece of leaf—it's not neat but it will do. She picks up the leaf in her mandibles and starts off down the track, following the pheromone trail she left for herself. Looking up, she sees a shiny, big-horned scarab beetle coming toward her, as if following her own trail. She drops the leaf, detours off the path, and freezes behind a log for thirty seconds. The beast passes. But she needn't have worried. The big rhinoceros-like horns, found only on male beetles, are designed for combat with other males of the same species: They joust with each other, and the loser is forced out of trees or away from prime feeding sites. Darwin and others presumed the male beetles fought over females, but the only advantage appears to be that females attracted to large-horned males are more likely to eat well, sharing prime feeding areas with them. Many of the 1,366 species of rhinoceros-beetle scarabs, subfamily Dynastinae, half of which are Neotropical, are almost as endangered in some tropical forests as the various mammal species of rhinoceroses in the savannahs of Africa and Asia. The main reason for their decline is probably loss of habitat; they need the large rotting logs found in primary lowland rain forests to lay their eggs and raise their larvae. As the forests have been cleared, these large logs disappeared, and so did beetles.

The leafcutter queen once again picks up the leaf and heads for home. Resealing the entrance, she jams the leaf down the dark tunnel. It drops to the bottom. She hoists it again, but it won't get through the entry to the chamber. She cuts it into threes, and this time it slides through. Inside the chamber, she tears one piece into even smaller pieces and chews them, then spits them out and squats over them, applying the fecal drops that make this such excellent fertilizer for the fungus. Then she wedges a few balls under the gray mass. It's awkward, and she ends up mashing a section of the garden with her clumsiness. When it comes to gardening, big queen leafcutters have no green thumbs, only left thumbs! Normally the ants that do this job are 825 times smaller, by weight, than the queen. But faced with the challenge to survive and with no specialists available, a queen must get resourceful. The next day she lays several more eggs in the chamber, and then with about twenty eggs as candidates for new workers, she brings the eggs over and places them squarely into the mass of spreading fungus. She is now only about a month away from having her first workforce. If she can just keep one garden going to feed the first brood, she can retire from gardening forever.

With leafcutter ants, the accepted wisdom is that the leafcutter queen, once she has laid her eggs and started her garden, never goes out again. Most new leafcutter queens, founding a colony, probably can manage to feed their gardens with their own fecal drops. Certainly, it's safer to remain inside, blocked off from predators. Any effort directed outside the nest is to make sure the next generation, her workers, are well fed. When the first workers appear, she will be able to relax and concentrate on egg laying and colony growth. It will be a good eight months, however, before she can make enough workers to raise enough food to support a few of the giant soldiers. Until then, the colony will be rather defenseless. Any workers that go out to forage will have to be sure to close the door. It is only when the soldiers appear that the leafcutter colony takes on the appearance of a fortress as well as a home, a proper castle that can afford to show its turrets and chimneys and allow its doors to be opened to the outside world. Until then, as the wars proceed overhead, to avoid the millions of jaws that patrol the forest floor, the young superorganism must lie low. There is also the matter of the new lab to be constructed nearby over the next couple of years. As the colony is situated now, it will be in clear view to any student visiting La Selva with access to a Berlese funnel or wanting to sift litter samples. Students often gather leaf litter samples from the forest floor and sift them through a funnel to sort species and measure diversity gradients among other things. The queen herself is safe from senior scientists and experienced researchers who know leafcutters well and know that a queen who has dropped her wings is a mated queen and should be left alone. But living outside the noncollecting zone and so close to the lab is a potential mistake that could lead to grief. In any case, from now on, her genes are telling her to remain down below.

Yet even as her genes tell her this, the failure of her fungus to grow leads her to make one last trip outside the colony. Her eggs have hatched into larvae. She feeds them some of the eggs she set aside, placing them one by one directly into the larvae's mouths and pushing them in. But the larvae are still hungry, devouring all the fungi within reach. She again leaves her wriggling larvae—the colony's future. Pushing away the dirt at her entrance hole, she is surprised by the brightness of the day. She climbs up out of her tunnel, the dirt clinging to her big shiny gaster, and giving her a bit of inadvertent camouflage. On a small scale, she resembles a burrowing rodent coming out of hiding.

Suddenly, she feels the ground pounding, at first from afar, then closer. A giant two-legged animal is walking through the forest. This animal is a scientist. As experienced as this leafcutter queen is, having survived a flood and an aborted nuptial flight, then having flown repeated nuptial flights over her world and founded a nest, she knows absolutely nothing about life of the two-legged variety. All she knows is her instinctive reaction to the foreign smell: to avoid it or bite it. She cannot conceive of an animal two hundred thousand times her size that would function like a colony unto itself. Her first encounter could be her last. Then she feels a pinching at her back and abdomen, as she is swept off her feet. Held delicately but firmly between the scientist's deft fingers, she smells a hundred strange odors, the smells of the scientist plus the many species of insects he has touched that day. The smells cloud her brain, overloading it. In any case, there is nothing she can do for the moment. The scientist holds her up to his jeweler's loupe. He touches her fat abdomen so gently, rubs it, then smells his finger. She has nervously leaked a little of her defense compound, but he cannot smell it. The scientist says something to himself. He notes her physogastric condition and feels a touch of sympathy. Then he puts her back down on the ground. The scientist is merely curious and admiring. It is an astonishing reprieve. There may be other crises later—battles and even wars in the life of her colony. Her colony is still in an awkward locale near the building site for the new lab and beneath a cecropia tree that may one day house a fierce aztec ant colony, but she has survived an encounter with a human.

She remains still for a long time. Then slowly she slips away to pick up a leaf piece and carry it over to her colony entrance. To see the queen struggling alone, it's hard to imagine that this single individual living in her tiny, modest hole in the ground may one day become the founder of a vast empire of gardens, with the roof stretching from twenty feet to fifty feet or more across, and at least twenty feet down in the ground, with hundreds of entrance and ventilation holes and up to 7 million inhabitants.

At least there are no swarm raiders around. According to one study in Panama, a leafcutter queen has a 50 percent chance of encountering army ants in 240 days. By the time the swarm raiders revisit the leafcutter's patch of ground, the colony should be large and ticking. She takes one last look at the daylight and disappears down below, sealing herself into the dark warmth for the rest of her life.

BY AIR ON royal wings, by land, under the ground, on sturdy legs climbing to the highest trees, and now by sea, the ants have spread out and have seeded the Earth in and around La Selva. The rains and flood postponed but did not prevent the mating of the swarm raiders and the leafcutters. It caused the leafcutter superorganism to move its nest—an operation in which it lost thousands of workers—but now the superorganism is arguably leaner and stronger, and its progeny is building its own reproductive line. With luck the new leafcutter queen will survive to raise a healthy new colony.

In terms of numbers, the leafcutters have suffered losses, but to a superorganism it is no more than shedding a few pounds, and now, having sent at least one queen out to found a new colony, there may be several million more leafcutters feeding leaves to fungus. The swarm raiders dividing in two will in time double their influence over the northern area of La Selva.

Over all of La Selva, no ant species has been driven out or made locally extinct by the flood. And two new species had floated in on the river from the middle elevations of central Costa Rica, tucked neatly inside a piece of leguminous fruit. It remains to be seen whether they will survive, whether they can find a niche. It would seem that few niches remain in the tropical lowland rain forest—that natural laboratory for coevolution and the most species-rich biome on Earth—yet it is impossible to predict how evolution by natural selection will play out. Even a few decades ago, who could have predicted just how diverse the rain forest is? Who knows the absolute limits of diversity?

At the same time, one of La Selva's ant species has been launched down the river on a raft—the little fire ants. They might not survive, but they have a chance. This ant species has fought battles on many shores, gained many a beachhead, and conquered. The questions remain: Where will the sea currents take them? Can they last the voyage? Where will they finally land? Who will meet them on the beach? Will they come out fighting, or will they be able to slip into a reduced animal fauna and take the time to build up resources and numbers before spreading out? Will they be able—like the leafcutters, the army ants, and the aztecs—to found a colony to pursue to completion the process that had started at La Selva so many days before in the season of so much mating.

Chapter Eight

Ant Wars

I shall next bring forward a scene still more astonishing, which at first, perhaps, you will be disposed to regard as the mere illusion of a lively imagination. What will you say when I tell you that certain ants are affirmed to sally forth from their nests on predatory expeditions, for the singular purpose of procuring slaves *to employ in their domestic business?*

—*William Kirby and William Spence*,
INTRODUCTION TO ENTOMOLOGY, *1815–1826*

The greatest enemies of ants are other ants, just as the greatest enemies of men are other men.

—*Auguste Forel*,
LES FOURMIS DE LA SUISSE, *1874*

The foreign policy aim of ants can be summed up as follows: restless aggression, territorial conquest, and genocidal annihilation of neighboring colonies whenever possible. If ants had nuclear weapons, they would probably end the world in a week.

—*Bert Hölldobler and Edward O. Wilson*,
JOURNEY TO THE ANTS, *1994*

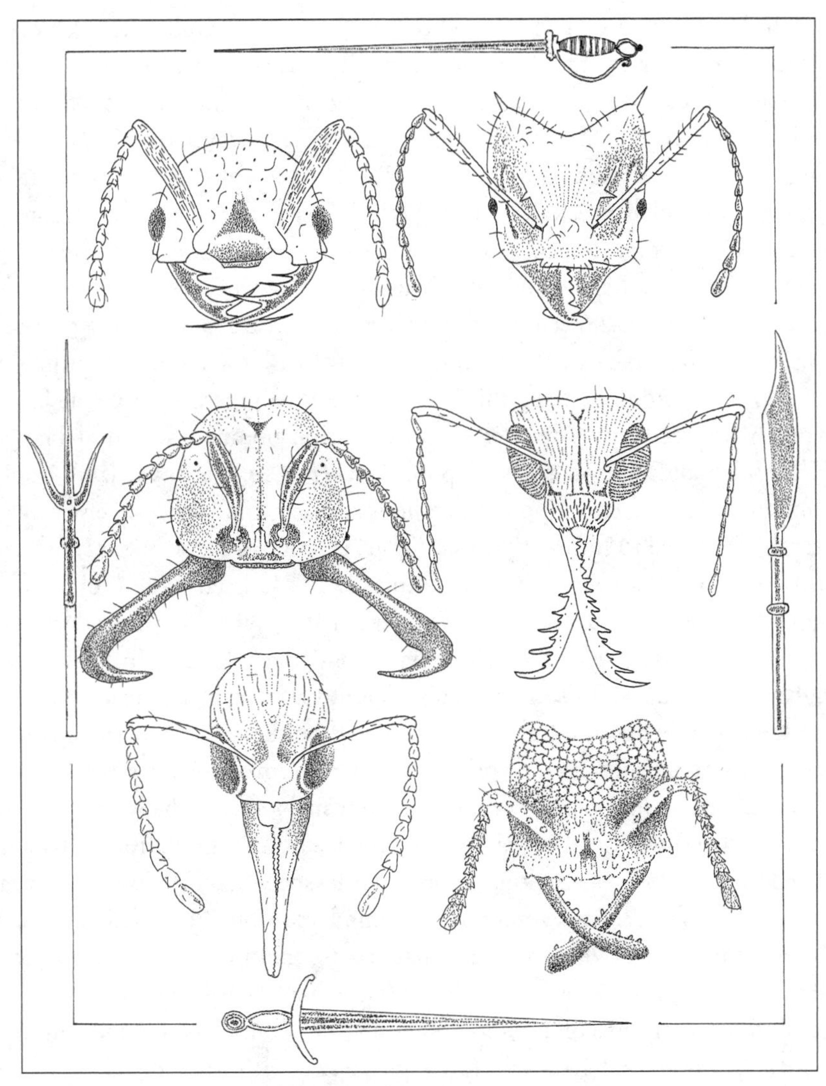

The Weapons of War. The mandibles of ants serve as offensive and defensive weapons, as well as prey-gathering and food-handling tools. This rogues' gallery of ants includes, clockwise from top left, the miracle ant, Thaumatomyrmex; *a leafcutter soldier,* Atta cephalotes; *a big-eyed* Myrmoteras *ant; the many-toothed* Mystrium *ant; the long-jawed* Harpegnathos saltator; *and a swarm raider army ant soldier,* Eciton burchelli.

A tropical rain forest field station is an unlikely place for debating the wars of the world. It is 1991, and visiting scientists and staff at the La Selva field station are gathered around after the day's fieldwork to share some beer and watch the TV or listen to the radio news about the Gulf War—Operation Desert Storm—as it unfolds in the Middle East, half a world away.

One of the visiting scientists participating in these informal debates is Ed Wilson, back for his fifth visit in eight years. He has traveled to La Selva alone this time. Wilson's sometime associate Bill Brown is grounded with cataracts on his eyes which precludes any fieldwork, certainly on animals as small as ants, and keeps him for now close to home and work at Cornell. Wilson's main field companion, Bert Hölldobler, with whom he had made three previous visits to La Selva, has recently left his professorship at Harvard for a professorship at the University of Würzburg. He's busy with the chores of settling into a new job and administering the $1 million German research prize he has just won. After nearly two decades of collaboration on ant research at Harvard, Hölldobler and Wilson recently published *The Ants,* an oversized, double-columned, 732-page, 7½-pound synthesis of their and everyone else's ant research; dedicated to future myrmecologists, it is one of those books most likely to be consulted well into the next century. It has been widely praised as a more than worthy successor to Wheeler's 1910 opus *Ants: Their Structure, Development and Behavior,* which had stood unchallenged in its scope for eighty years. And as Wilson shows up alone at La Selva, he is privately celebrating their winning of the Pulitzer Prize. Their book was a scholarly work, the kind that rarely wins literary prizes; it had a German coauthor, *and* it was a book on ants—all of which speaks of the unusual merit of the work. But more than this, everywhere he walks, with the chance of reflection afforded by a solo expedition, Wilson is thinking about all the species they have studied and written about here. He wouldn't even mind meeting the fabled miracle ant *Thaumatomyrmex,* which he had searched for before, though colleagues in Brazil recently found one after Wilson had issued a challenge to the younger generation of myrmecologists as something he would like to find out about before "going to that great rain forest in the sky." It turned out that this ant's bizarre pitchforklike mandibles are adapted to catching and stripping the skin off tiny polyxenid centipedes.

But mostly this trip is conducted in the long shadow of war. Seeing ant-sized humans on a small TV—and all the talk of smart bombs and the threat of chemical warfare in the deserts of the Middle East—makes Wilson think anew about how ant wars are conducted. There are the battlegrounds of Wilson's youth, the American South, as well as some of his favorite study sites that he had written about with Hölldobler, such as the South Pacific islands, the Australian outback, the Sonoran deserts, the Old World of Europe, and La Selva, Costa Rica. All ant wars are strictly ground—as opposed to air or sea—battles, though trees, underground bunkers, tunnels, and even bridges figure prominently. These social insects, which have lived on the Earth and have been fighting each other for at least 100 million years longer than humans, have developed some fascinating refinements in the so-called art of war.

"Ants are by far the most warlike of all social organisms," says Wilson. "They conduct wars which dwarf those of human beings." But many ant species expend considerable energy to make peace, or at least to avoid conflict. Wilson and Hölldobler have researched the astonishing range of offensive and defensive behavior performed in the course of ant wars. As Wilson says, the average person wouldn't believe the stuff going on between the cracks of the suburban sidewalks of America—never mind the deserts and rain forests of the world. And ants were going to war long, long before the ancient Sumerians set up the headquarters for their empire in the city of Babylon and the first recorded human wars were fought, almost six thousand years ago.

As Wilson walks the paths of La Selva and carries on with his program of field research in an always too brief hiatus from teaching classes and managing graduate students, the affairs of ants and humans seem more relevant than ever. Wilson and his ant buddies come here partly to get away from the world. Restoring their enthusiasm for life is a matter of walking through the jungle for a few hours until they find something new and interesting to watch and study. But this past year, world events have intruded more than usual. There was new optimism about world peace in the late 1980s as the Berlin Wall fell and the Iron Curtain parted. Now in the aftermath of the Gulf War, with ethnic wars simmering in eastern Europe, everyone is thinking about the inevitability of conflict. And so Wilson, Hölldobler, and others are looking with fresh eyes at the societies of ants, chimpanzees, and other organisms for clues to the nature and biological basis of war. The lives of ants and research into them are taking on new meaning and vividness.

On one of the first days in the field Wilson finds himself, as he has done at least once on every Costa Rican expedition, in the midst of the army ant swarm raiders, surrounded by hundreds of thousands of ants. It is the southern swarm raider colony formed several years earlier when the mother colony divided. All the original workers that rebelled against their mother are long gone—save for the sister who became queen and remains mistress of the expanding hordes. The superorganism is now near the peak of its size and ferocity. It has acquired hundreds of thousands of camp followers featuring mites, flies, beetles, and birds, in descending order of numbers, an assemblage that in total biomass exceeds the army ants themselves. Over the past four years the swarm raider superorganism covered some fifty miles of shifting trails crisscrossing the northeast corner of La Selva. With an estimated minimum of thirty thousand food items eaten per raid, this single superorganism, since it started four years ago, has consumed more than 19 million insects and other prey, mostly small arthropods. It ate its way through some of the most diverse and interesting small fauna in the world. It even consumed species of ants and other arthropods unknown to science. Surprisingly, it avoided all the other swarm raider colonies in the area, including the northern daughter colony that it had separated from. Soon, of course, it will be dividing itself, but for now it raids with an awe-inspiring drive, displaying its supremacy wherever it goes at La Selva, flexing its superorganism muscles and running through the jungle with a swagger. Early naturalists likened the army ant raids to the ultimate example of war. Look at the language used to describe them and their activities: army ants, swarm raiders, bivouacs, and booty caches. This is the language of battle. But is this really war?

The brown river of bubbling, boiling life spills out across the jungle. The humming swarm of attacking ants is a huge tropical monster on the loose. And it is flowing toward the senior myrmecologist standing beside the trail. He looks up and across the broad fan of the swarm front. It is in the process of expanding from thirty feet wide to its full width of about fifty feet. In a moment it will be upon him, and if he doesn't move, the soldiers as well as the workers in the first rank will be advancing up his legs, snapping at the fresh smell of living tissue—in this case, human flesh. Wilson is unconcerned because he can shake his legs, stamp his feet, or simply move out of the way. But if he were to stay put for a few hours as a sort of endurance contest—such as certain South American native tribes employed as a rite of manhood or as a punishment—thousands of mouths would visit his flesh. Long before that, though, the bites and stings would convince him of the folly of his experiment, and he would flee.

To Wilson, his good eye sharply focused and his mind attuned to life at this scale, the lethal, ice-tong mandibles of the soldiers positively gleam in the light. He cannot help but admire the long, sturdy legs or fail to notice the tiny mites that adorn the bodies, legs, mandibles, and heads of the swarm raider workers. Against the reddish black backgrounds of their bodies, the mites are dark spots. He also takes note of the many beetles, some colored like the ants in the center; the odd ones on the outside fringes of the swarm raid stand out as they dart in and out to grab pieces of meat thrown up by the army ants' advance. He sees a few silverfish that have become myrmecophiles. He sees the flies buzzing overhead. If he's lucky, he notices a bird or two.

To the myrmecologist, the experience of an army ant raid is largely visual. (Wilson himself is hard of hearing and can't pick up the higher registers of sound.) But there are as many other ways of perceiving animals such as army ants as there are senses or combinations of senses used by particular groups of animals and the scientists who study them. The dipterist, or fly specialist, *hears* his flies, and even at some distance this can lead him to a swarm raid. The ornithologist hears the shrieks and scoldings of the antbirds that follow the swarm raiders and feed on the flushed-out, escaping beetles and other small insect prey. There can be up to fifty species of so-called antbirds in a single flock trailing an army ant superorganism, though the antbirds that follow the swarm raiders in Costa Rica are usually composed of considerably fewer species. Most of the birds are rather small and dark, and difficult to see. Some have niches—on underbrush or on open foliage—but most share the same or overlapping territories. No wonder skirmishes develop from time to time. And all, separate and together, are noisy.

But to most other ants and insects, the visual and audio channels are nothing compared to the smell. The primary sensory experience of most ants toward army ants is an overpowering foul scent. Each army ant superorganism has its own slight variation of this fetid odor. Everywhere it goes, it leaves a smell that can take days or even weeks to fade. For the most part, ants do not even distinguish the beetles, mites, and other guests from the swarm raiders they travel with. The guests, to earn their acceptance, have had to assume the same or similar fetid odor. And that smell, to most of the rest of the jungle, means only one thing: "Danger. Flee now."

Wilson has come to enjoy the hint of this smell—which he describes more as "musky" than fetid, though he can't distinguish the subtleties of the many pheromone blends that ants routinely employ. Enthusiastic about the prospect of watching the ants and being surrounded by them, he walks toward the charging army. He crouches down to get an ant-size view of the proceedings. For a few moments he is a boy again, lying on the ground, watching the extraordinary battles of ants—much better than any toy soldiers. Indeed, it looks like an advancing army. From the ground the huge soldiers with red-orange headgear and the long ice-tong mandibles that almost drag on the ground advance like infantry in the front line. The soldiers also have a black alitrunk, or midsection, and a reddish orange gaster, giving them the appearance of wearing full dress uniform with chitinous armor. Behind them are the submajors who carry booty back to the colony and the media workers who do a little of everything. They have black heads, short mandibles, and reddish orange gasters. In the midst of the advancing army are the beetle myrmecophiles, colored similar to the army ant workers but looking more like miniature armored tanks. Off to the sides of the main platoons, looking a little disheveled and disorganized, are other beetles and assorted arthropods in motley uniforms and larger tank armor. Just above the battlefield, the thousands of flies are like Spitfire warplanes or guided missiles, while the big antbirds almost resemble giant bombers or transport planes. Wilson can now just make out the insistent calls of the antbirds and the whining flies that sound almost like the humming advance of war machinery. And then come the smells of battle again, filling the scientist's nostrils more intensely than before—the fetid smell of army ants mingled with the smells from thousands of insect lives being snuffed out, the smell from the corpses being carried underneath individual ants' bodies. The unholy smell of combat.

And at the end of the raid there is an unlikely scene—something Wilson and Hölldobler first witnessed a few years ago at La Selva. Limping at the rear is a contingent of older and obviously infirm workers, some with broken legs, others with bruised, misshapen exoskeletons, or missing antennae. Behind them are various generalist rove beetles who dare to jump at random into the swarm and try to catch one of these infirm workers, grasping a leg or the tip of a gaster and hauling down the big, long-legged animal. They bite the ant savagely and drag it off. In some cases two beetles attack a single old soldier, coordinating their effort and sharing the spoils. These unantlike beetles could never gain admission to the swarm raider colony, so they work the fringes in a dangerous cat-and-mouse assault, focusing on the sick, injured, or old swarm raider workers.

Thus the beetles nibble away at the swarm raider superorganism, showing that the contest is not entirely one-sided. But is this fierce, awesome creature —the swarm raider superorganism—literally "at war" as it would seem? Wheeler called the army ants "the Huns and Tatars of the insect world," which implies great battles and an appetite for war. Yet he understood that what was happening was feeding behavior, not war, adding that the army ants are "filled with an insatiable carnivorous appetite." What seems war to the many ravaged creatures on the forest floor, in the trees, and in the underbrush is really predatory behavior, albeit on a grand scale. Rettenmeyer described the superorganism best as "comparable to a wolf pack, but with fifty thousand miniature wolves . . . the epitome of group predation." And the beetles that prey on the sick and old army ant workers are also following a predator-prey pattern of behavior common in large mammals. Competing over a good meal could drive a species to go to war, but are army ants ever truly warlike? If army ants do not go to war, which ants are warlike? Are ants, in fact, capable of war?

Wilson stands up and brushes off the swarm raiders that start advancing up his trousers. Try as they might, the swarm raiders cannot subdue the curator of entomology from Harvard's Museum of Comparative Zoology and carry him back to their booty cache. He turns his attention toward a familiar spot near the clearing that opens up to the new lab site. He notices a big bullet ant, a solitary forager leisurely probing in the open along tree branches and large leaves, her hefty coal black body standing out clearly against the bright green leaves. Here I am, she seems to be saying. Don't mess with me! The centerpiece of the clearing is a nearly mature cecropia tree trying to secure its place in the canopy. Wilson sees two aztec ants scuffling before becoming locked in combat. He smells their mildly repugnant odor, which resembles butyric acid, the acids found in butter and other animal milk fats sometimes used in the making of disinfectants and drugs. The cecropia houses two large colonies of aztec ants in its twin hollow forked trunk. This scuffle is one skirmish among many that will soon lead to what can truly be called all-out war.

At the foot of the cecropia tree is the large leafcutter colony. Since Wilson was last here, the leafcutter queen who survived the flood founded a colony that has grown to almost 5 million workers. While the underground agricultural collective has kept busy with its many and various farm chores, the aztec ant colonies in the cecropia tree have fought each other. The cecropia tree and the individual aztec colonies have kept growing in size, but the number of colonies has been declining. One by one they have been killing each other in a series of battles during an intermittent five-year war. Now only two colonies are left—an eastern and a western fork colony. In the time the tree has quadrupled its height, ant numbers have expanded perhaps one hundred times. Soldiers of both colonies—east and west—patrol the DMZ between them, an area four feet up from the base of the tree where the trunk branches in two. It's the no-ant's-land between the two last aztec superorganisms. As for the leafcutters, they do not seek and destroy other ants. Yet because of the warrior aztecs above in the cecropia tree, they must keep soldiers posted along the northern stretch of the colony roof, ten feet away from the trunk of the cecropia tree.

Wilson is careful not to brush the cecropia tree, which would send these true warmongers on a rampage. The final battle for the control of this tree, now counting down, with skirmishes increasing every day, will dwarf all other previous contests. In this plot of La Selva, it will be the final war for control of the aztec world.

WHAT IS WAR? The American Heritage dictionary defines war as "open, armed, often prolonged conflict carried on between nations, states, or parties." It is therefore possible only in animals that form groups—such as humans, monkeys, and ants. But only humans use tools for war—arms or weapons—though there is one tantalizing example of tool use as a possible offensive strategy in a desert ant species. Finally, to qualify as war, it should be combat against one's own or a closely related species. Army ants direct their raiding activities against other species—non–army ants—and the victims are individuals, not groups. It's predatory behavior, not war. The same goes for ponerine ants that hunt alone—seizing other species of ants or arthropods. And individual ants in isolated incidents of one-to-one combat are fights, not wars. Other ants, like the aztecs, do battle against other colonies of their own or closely related species, fighting to the death. Like humans, these and certain other species of ants go to war over territory, food, or other resources. But there are also many ant species who have evolved strategies to avoid combat—everything from camouflage to specialized techniques of defense. They sometimes have big bodies with armor or spines. They may use their own plug-shaped heads to block off their colonies. Defensive strategies are even more numerous and every bit as clever as the offensive techniques.

The real origin of ant wars is long lost in geological history, but they may have started as plunder. Many ant wars feature the smashing of other nests, the taking of food, and even seizing slaves, which for some ant species is a full-time occupation.

Ant "weapons" are mainly derived from two common features of ant anatomy: the sharp-toothed mandibles and the chemical-producing glands. The importance of the mandibles for ants is on the order of the opposable thumb in humans. The mandibles of primitive ants are two short blades extending out from the mouth with a row of sharp teeth on the inside edges used for gripping and cutting. The primitive ancestral ant, *Sphecomyrma,* that Wilson, Brown, and Carpenter identified in the amber, had two broad teeth protruding from simple mandibles. Mandibles may have originated as digging tools in pre-ant ancestors. These may have evolved into prey-seizing and -carrying, as well as digging, tools in primitive ants, and that is still their primary function today in both primitive and advanced subfamilies. They carry not only food but dirt, water, and other nestmates, as needed, especially eggs and larvae that must be moved to food or for temperature regulation. But in many ants the mandibles have evolved to look like wire cutters, pruning shears, pincers, ice tongs, or pairs of sabers, swords, sickles, or pitchforks, or other shapes. As ants have spread out and taken over much of the Earth, then increased their density in many areas, the competition from colonies of the same or similar species for food and territory has helped produce this array of weaponry. In Barry Bolton's superb *Identification Guide to the Ant Genera of the World,* the scanning electron microscope photographs of the three hundred or so genera of ants make for a rogues' gallery of bizarre creatures with odd-shaped heads that are hairy, creased, or bald, with big eyes or small. But more than anything it's the mandibles that, on page after page, give one the feeling of looking through a medieval armaments catalogue. These are animals that no unsuspecting creature would like to meet on a dark night or even at high noon.

Even more important for warfare than the mandibles is the battery of exocrine glands that produce the mix of chemicals the ants use for their communication system. Some of these chemicals are used for chemical warfare, which Hölldobler and Wilson say is the main way ants fight. In humans, chemical warfare is nasty—from mace to mustard gas. In ants, it is no less unpleasant—for other insects, and occasionally for humans. The most painful ant-inflicted wounds come from stings—the injection of painful venom, such as from the big bala ant. The aculeate hymenopterans, such as bees, paper wasps, and hornets, have stings, as did the ancestral ant. Most ants still carry stings. In the ant subfamilies, Formicinae and Dolichoderinae, as well as in some myrmicines, the sting has become vestigial; instead of a sting, the various formicine ants carry and use formic acid. Chemical warfare as waged by the Formicines consists of bites followed swiftly by a squirt of formic acid in the wound. It's not the bite that hurts; it's the acid. Thus, chemicals can be used not only as a means of talking to colony members to recruit them to food, but to make alarm calls to call them into battle, to kill or immobilize prey, and to attack and kill other ants in battle. By adapting their chemical weapons into a superb pheromone communication system, warlike ant species have evolved to wage ever more sophisticated wars, calling out new units as needed to carry out specific tasks and to deliver their chemical weapons.

In the smaller, primitive colonies, all the workers may handle defense or offense. In the more advanced, highly social ants, there are a number of specialized castes. To protect the interests of large colonies such as the leafcutters, army ants, weaver ants, and the aztecs, the major worker or soldier caste contains warrior ants, each typically with a bulky body, a large head with oversize mandibles, and a powerful bite and/or sting.

The soldiers often look like a completely different species from their smaller nestmates. As Wheeler first reported, the enlargement of adductor muscles in the head gives increased crushing and shearing power to the mandibles and even affects the overall size and shape of the head, making it large and heart-shaped. Of course, these ant warriors are all females. Males, though often larger than the workers, are inadequate at colony defense and often helpless to defend even themselves. Reproduction is their only role in ant society.

The older female workers, in some species such as weaver ants, are usually stationed near the periphery of the nest and are the first to be attacked and killed when wars or other skirmishes occur. These older females are non-reproductives; but even the all-female worker forces, which often find themselves on the battlefield, are non-reproductives. In humans, the young males are the prime reproductives, and now with equal opportunity armies in the world, young women also go to war. From a strictly biological point of view, this policy would seem dubious for a long-term survival strategy. As Wilson enjoys saying to the gathered field-workers at La Selva in the aftermath of the Gulf War, humans send their prime youth to war, but ants send their old ladies.

The basic ant-fighting techniques, as compiled by Hölldobler and Wilson, feature various offensive and defensive strategies. Among those that can be classified as mostly offensive techniques are the various ant weapons of mandibular and chemical warfare, including saber, hatchet, and trap mandibles, and various stings, sprays, and secretions. Some species use these for defensive purposes as well or instead, but mainly they're for offense. This is the weapons inventory that evolution has offered various ant species—their peculiar "weapons" catalogue:

Mandibular techniques: sabers, hooks, hatchets, and traps

- Hatchet-shaped mandibles can slice into the chitinous exoskeleton or chop off the legs of an opponent. Ants that fight by shearing include many *Pheidole,* leafcutter, and certain carpenter ants.

- Saber- or hook-shaped mandibles with pointed tips that turn in at the end are sported by various soldier ants such as the swarm-raiding and column-raiding army ants, and some of the notorious slavemaking species. When an enemy ant gets its head or body caught inside these jaws, the result is instant death as the soldier or other attacking ant simply squeezes, puncturing the integument.

- Trap mandibles, as used by the dacetine tribe and some of the ponerines, consist of elongated mandibles that snap shut convulsively, impaling prey or enemies on the sharp teeth at or near the tip.

Chemical warfare: stings, sprays, and secretions

- Almost all ants use some chemical warfare, ranging from venomous stings to poisonous droplets and secretions. It is their preferred method of combat. Stings are widespread throughout the ants especially in the primitive groups but also in some of the more advanced ants. The big ponerines such as bullet ants are famous for using their venomous stings, but the army ants and even the leafcutters also employ stings. Ants with reduced or missing stings tend to use sprays and drops from other glands. Many of the tree-based ants such as the aztecs are stingless ants that bite their victims and then spray or rub poison in the wound. Stings and sprays are sometimes used as a defensive maneuver but are mainly offensive.

- As part of chemical warfare, many ants also employ pungent (as opposed to poisonous) secretions from their various exocrine glands. These secretions are sprayed or wiped on the enemy, or sometimes are simply allowed to evaporate from the ants' own body surface.

TECHNIQUES THAT ARE exclusively defensive include the use of size, armor, spines, and head grooves; bouncing, blocking, and sealing off the nest; the use of warning coloration or camouflage; building protective tunnels or walls; staging evasive movements or full retreat. Ants are among the most violent social animals, yet they have evolved a surprisingly extensive list of defensive strategies to avoid conflict at all cost. These strategies can be classified as follows:

Defense by size, armor, spines, and head grooves

- Size alone can be a defensive technique. Some ants are comparatively large, such as the bullet ants. Size, combined with a powerful sting, provides effective defense against most other animals. Also small size, the "pygmy defense," has been mentioned informally by myrmecologists and may be an evolutionary strategy having to do with defense. The size variation in ants is extraordinarily wide, with an entire colony of the smallest known ants able to fit inside the head of the largest ant.

- Armor is another way ants avoid being eaten. This strategy is used by various ponerine and other ants for defense when foraging for prey. Often combined with armor are spines typically found on the thorax, head, and petiole of leafcutters and a wide variety of other ants in most subfamilies. These spines make the ants less palatable to predators, especially if the ants are large and have solid armor. One ant, the golden-haired *Polyrhachis,* has lethal-looking spines protruding like hooklike thorns from its waist.

- Head grooves to hold part or all of the folded-back antennae are good protective anatomical features in many of the myrmicine species and a few ponerines, including the bullet ant.

Defense by bouncing, blocking, closing off the nest, or building tunnels and walls

- Bouncing is a specialized form of defense used in at least two ponerine ant species. The soldiers of the Australian dacetine *(Orectognathus versicolor)* have long, blunt mandibles. When an alien ant enters the nest, the soldiers spread their mandibles wide. As the intruder approaches close enough to touch, the soldier snaps her mandibles closed on the enemy. The enemy ant is pinched out, as Hölldobler and Wilson describe it, "like a slippery seed pressed hard between the fingers. In this nonlethal technique, an enemy ant can be shot four inches into the air and effectively deterred from further investigation." In a similar way, certain trap jaw ants (*Odontomachus*) bounce invaders from the entrances to their nests.

- Blocking behavior—usually with the head but sometimes the abdomen or gaster—is a specialized defensive technique. Typically, the nest entrances are blocked by the big saucer- or cork-shaped, flattened head of a soldier. The Neotropical cork-headed ants, *Zacryptocerus,* as well as the *Colobopsis* ants from Europe and North America, have soldiers that stay in the nest and use their head and upper body as a sort of living door. Workers of the colony who have the chemical-sensory password are allowed in. The soldiers simply step back into the nest and let them pass. In some cases the top of the head carries dirt to give it some camouflage when pressed into the hole and visible from the outside. If the hole is larger than one head, in some species two or more soldiers will join together to block the entry. Reverse blocking, when found, is practiced by all workers rather than just a specialized caste. One species of *Pheidole* ant is suspected to use its gaster, and a *Proceratium* species employs the rear surface of its abdomen—both pressed against the nest openings to keep out intruders.

- Closing off the nest is another strategy. The harvester ant uses soil to close the nest at sunset or when attacked by other ants.

- Building tunnels or walls to cover trails or food sites with soil and carton material helps protect foraging workers and good food sources. Wheeler was the first to report that various ants do this, such as the acrobat ants, *Crematogaster.*

Defense by evasive movements and full retreat

- Evasive movements can also be a good defensive strategy. Besides simply running and hiding under leaves, which some ponerine and other ants do, the trap jaw ant is known for being able to jump and bounce, while other species will deliberately fall from branches or leaves often high off the ground.

- Full and fast retreat, completely absconding from the nest, is the best defense for certain weaver, *Pheidole,* harvester, and fire ants. Wilson studied the rapid, well-organized evacuation of the *Pheidole* ants in response to water, but certain ants also respond in the same way to formidable enemies. Fast action can be crucial to a colony's survival.

Defense by "professional" soldiers

- Having a soldier caste is in itself part of some ants' defensive strategy. Instead of fighting, a colony or sometimes even an individual worker can summon the soldiers to defend the nest or workers looking for food, or to protect food that has been found. In their role as the protective caste for the colony, soldiers employ four main fighting techniques: shearing, piercing, blocking, and bouncing. These are some of the same key techniques used by all fighting ants and the same ones used in the course of ordinary hunting, such as when the army ants go swarm raiding or when the leafcutters defend the colony or try to protect the foragers working outside the nest. But other ants, truly warlike ants, use these techniques to fight their own and other closely related species.

Defense by warning coloration and camouflage

- Warning coloration is not employed by ants to the same extent as other tropical species, such as various poisonous frogs, but certain ants *(Pseudomyrmex, Myrmecia,* and *Macromischa)* are red and black or metallic blue and green. Typically, these are bold ants that forage in the open and have strong stings. The bright colors may warn vertebrate predators to beware of eating them.

- Using cryptic coloration—camouflage—some ants "play dead." With so many aggressive ants it is inevitable that some species would take alternative evolutionary routes to fierce aggression. There are a number of small members of the Attini tribe of gardening ants, as well as other small myrmicines, that try playing dead so they won't be noticed. But evasion by camouflage is taken to an extreme by the cryptobiotic ant *(Basiceros manni).*

When Wilson and Hölldobler first encountered a cryptobiotic ant colony at La Selva in 1985, they assumed the species was extremely rare. But once they knew how to look for it—by noting the white larvae and pupae and then staring hard until the colony suddenly appears—they started seeing it on every visit to La Selva. One aspect of this ant's camouflage is its extreme sluggishness. It creeps about the forest floor, an unlikely huntress, and the only thing fast about it is the snap of its jaws when it locates and manages to approach close enough to an unsuspecting insect. The prey is surprised because of the slowness of the approach and the fact that the ant can remain absolutely still for minutes at a time, without even moving the antennae. But the story is even more bizarre. The source of the camouflage, as revealed by scanning electron microscope photographs, is dirt. Most ants spend a considerable portion of their time licking each other, cleaning their own or other ants' bodies, but cryptobiotic ants devote little time to personal grooming and wear the equivalent of "khaki battle fatigues" composed of dirt particles. The photos reveal two kinds of hairs on the ants' bodies—some clublike and the others like feathers—that actually collect and keep dirt on the ants as a permanent feature of their anatomy. All of this prompted the two scientists to call these the slowest, dirtiest ants in the world. It is a demonstration of how evolution helps create such a diversity of life that almost anything is possible. Many ants are quick and aggressive, and almost all are obsessively clean. But the cryptobiotic ants go the other way and have found in it a successful defensive and offensive strategy.

Like football players, some ants may be good blockers on defense but clumsy at offense. When Wilson raised the alarm in *Colobopsis* colonies, he found that the soldiers moved to the nest entrances, even filling holes that had been unattended prior to the alarm reaction. On the other hand, the soldiers were inept at combat. When twigs containing these colonies were first broken open, both minor workers and soldiers rushed out. Many attacked any accessible alien object, such as the observer's hand or a bit of cloth offered to them, biting it and spraying it with formic acid. The same response is obtained in the laboratory by permitting fire ant workers *(Solenopsis invicta)* to invade the nests. And concerning Australian carpenter ants, Hölldobler reported that when a stick is touched to the nest entrance, the guarding soldier attacks it with her powerful mandibles. When the stick is removed, the soldier usually does not release her grip and is pulled out to what would be her death. Immediately, however, another soldier takes her place at the entrance. In this way these carpenter ants can hold their own through a stolid, altruistic defense. And for these ants, steadfast defense rather than going head-to-head in offensive combat is the only way they can survive in the territories of the dominant *Iridomyrmex* meat ants.

Defense and offense, therefore, are much more than just a matter of weaponry. The ways of ant wars are as various as the thousands of species that wage it. In the territorial wars of the pavement ants *(Tetramorium caespitum)* along much of the American eastern seaboard, warring colonies of the same species mass along their borders and bite and kill each other until one colony is driven away or the queen is executed. These summer-long wars—first described in lurid detail by American myrmecologist the Reverend Henry C. McCook—are basic primitive wars, no fancy techniques or strategy. Might makes right. The biggest colony almost always wins.

These wars may be protracted affairs. Some of the harvester ants, the industrious ants praised long ago in the sixth chapter of Proverbs in the Bible, actually turn out to be some of the most warlike, waging bloody campaigns that have, according to two authors, lasted twenty-one and forty-six days. These rugged ants are superbly adapted to desert life, able to survive, according to the monitoring of a California colony, for up to twelve successive years of severe drought.

The mound-building wood ants of Europe fight cannibal wars, particularly in the early spring when lots of food is needed to get a colony going. After battle, the colony carries its neighbors back to the nest and eats them. Ants commonly eat other ants, but cannibalism—eating members of one's own species—is much less common.

The woodland ants, *Pheidole dentata,* use preemptive attacks to keep at bay their much larger, dominant neighbors, the notorious imported fire ant, *Solenopsis invicta.* In confined laboratory situations, Wilson watched the fire ants defeat the *Pheidole* ants again and again. But in the pine woods of the southern United States, the *Pheidole* ants manage to live in the same areas as the fire ants. The *Pheidole* ants have a responsive soldier caste with powerful mandibles that can snip off the limbs of the enemy fire ant scouts. As part of a dedicated strategy, the entire colony reacts instantly and with full force when it sees one or even a few fire ant scouts. They are eliminated so quickly that no message even gets back to the fire ant colony. With the *Pheidole* ants taking out all the scouts that venture near their territory, the fire ant colony is kept blind. And if their colonies are somehow found and attacked, the *Pheidole* soldiers block the fire ants' approach, giving up their lives and dying from the stings of the venomous fire ant while the rest of the colony evacuates, moving the nest.

The foot soldiers of both humans and ants are expected to be fodder for war. Compared to human soldiers, however, ants are readier to sacrifice their lives on a daily basis for the colony—to the point of committing suicide to defend their sisters. The most dramatic example occurs in the Malaysian forests. There, German entomologists Ulrich and Eleanore Maschwitz found carpenter ants (*Camponotus saundersi*) that are programmed by their peculiar evolution to be walking bombs. They have two oversized glands that run from head to gaster, each filled with poison. When these ants are cornered or attacked, they contract their abdominal muscles violently, their bodies burst open, and messy, poisonous secretions are sprayed all over, killing or disabling their opponent and any other ants in close proximity. Like the human suicide car bombers, the ant bombers die instantly.

The weaver ants, who fight fierce arboreal wars in the Old World tropics of Asia, Australia, and Africa, also employ their sophisticated pheromone communication system for war. The Allies in the Gulf War relied on an excellent communication system to give them advantages in battle, and so do the weaver ants. One pheromone can summon workers to a nearby enemy; another sends them to an enemy some distance away. Weaver ants can issue chemical mixtures containing four-part messages that first get the ants' attention, then cause them to become alarmed and to search for the danger, then to come closer and bite anything foreign, followed finally by the aggressive push to attack and bite. Staging African weaver ant colony wars in potted lemon trees in the laboratory, Hölldobler and Wilson found that the ants defecated all around their nests, spreading the mess all over their territory, as far as the scientists would allow it to extend. In one of these areas, the scientists let wars be fought, and the winner was always the ant colony that was allowed to go in and mark the area first. Home field advantage can be a deciding factor for ants as well as humans.

A relative of the trap jaw ants *(Odontomachus)*, which Wilson had studied as a teenager in the 1940s, literally catapulted its way to fame at La Selva during one of Hölldobler's and Wilson's visits. Hölldobler was starting to examine a flimsy thatched ant nest in a tree when all of a sudden some twenty ants snapped their mandibles and sailed nearly a foot and a half through the air and landed on Hölldobler. They immediately started stinging him. For all the ants knew, he was a large mammal attacker looking for succulent larvae. He stepped back. It was the closest thing he'd ever seen to an air strike by ants, and Hölldobler decided to make a study of them. These ants are able to catapult themselves through the air by pointing their heads downward against a hard surface and simply snapping their mandibles closed with astonishing velocity. The catapult is only a secondary use of their jaws; they are mainly used to capture prey. Like a dacetine, the trap jaw ant lies in wait or creeps up slowly, its mandibles wide open. Then when sensitive hairs around its mouth are touched, the jaws snap shut to impale, kill, or stun the hapless insect prey. Hölldobler was so impressed with the speed of the jaw closing that, with two associates, he decided to try to film it on an ultra-high-speed camera at three thousand frames a second. It turns out to be the fastest recorded anatomical structure in the animal kingdom, faster than the jump of a springtail or the escape of a cockroach. It takes but one-three-thousandth to one-one-thousandth of a second for that jaw to close. The lethal blades of the fastest jaws in the west travel at a speed of twenty-eight feet per second—a galloping twenty miles an hour.

Like humans, ants are so clever, so successful, so dominant that their density is often high, or as Wilson and Hölldobler put it, "They're often living within shoving distance of one another." As the density and dominance increases, so do the wars. And the weapons, whether missiles or mandibles, become ever more sophisticated and keep evolving.

Humans, because of their ability to make tools, have taken wars to a new and frightening level. Modern human wars, with nuclear and chemical weapons, terrorist activities, and an emphasis on technology, are in a different category altogether from the wars of ants. And it's a good thing. If ants were to evolve much more damaging weaponry, the days of the Earth might truly be numbered.

FAR FROM LA SELVA, when Bert Hölldobler was a high school student in the early 1950s growing up in Ochsenfurt, near Würzburg, Germany, he used to stop and watch ants in the nearby Bavarian woods as they carried on their incessant daily battles. He was already something of an ant fanatic, gathering and keeping all the local species in homemade formicaries, sandwiched between his collections of butterflies and beetles. He also kept a pet flea in a vial and fed it blood from his arm. With an extraordinary eye for detail, he sketched and later painted the ants and other insects under his care, highlighting their special anatomical traits. Most of this could be dismissed as a boyhood obsession with bugs, but Hölldobler knew from the start that insects could be a career or at least a sustaining, lifelong passion, though, for a while, painting in itself held almost equal sway. His father was the entomologist Karl Hölldobler, a medical doctor who did research on the side and whose papers on the wasps and beetles that reside as guests in ant nests are still cited. Accompanying his father on walks through the nearby woods, the boy cultivated the naturalist temperament.

One ant that he loved to watch, the so-called amazon ant *(Polyergus rufescens),* would take ant slaves (various *Formica* species) as part of its warlike strategies. The amazons are large, robust ants, mostly shiny red in color; only the big-eyed, winged amazon males are black. The amazon females, the workers and queen, have black beady eyes that stand out against their red faces. Their smooth, polished, curved mandibles look like sickles, with sharply pointed, lethal-looking ends.

The slaves, which outnumber the warrior amazons at least five to one, are almost the same size as their captors but are all black. Some lab researchers daub paint on the backs or gasters of ants they are studying to tell them apart, but it's not necessary with these ants.

The amazon worker ants—if you can call them "workers"—stand around the colony doing nothing while their slaves keep the place clean, excavate new nest chambers as needed, go out to get food, and manage the care and feeding of the brood. The nest architecture follows the slaves' own design. The amazons are, in fact, incapable of even picking up their own food. Their tongues are so short that they cannot eat without assistance and must rely on the slaves to regurgitate liquid food right into their mouths. About all their tongues are good for is grooming—as they spit and polish each other's armor. Fortunately for the amazons, their slaves are mostly devoted attendants. If the colony must be moved for some reason, the slaves carry the amazons to the new nest. Indeed, new nests can be founded only with the aid of slaves.

While most of the amazons loaf around in the nest, at least one amazon is usually at work, scouting the nearby countryside for new nests of prospective slaves. At a signal from a returning scout, typically in the afternoon, the amazon workers suddenly come alive and leap into action. Sometimes the slaves try to restrain them from leaving the nest, but they push out. The transformation in their behavior is immediate and complete. As they bolt out of the nest, these warriors turn ever brasher and bolder, and charge in a column toward a *Formica* nest up to 250 feet or more away. They seem to know just where to go, which entrances to storm. The *Formica* ants work hard to wall up the nest, but the amazons charge through and head right for the brood chambers, relying on surprise attack and decisive action. One by one they pick up the pupae and carry them away. Inevitably, some of the workers guarding the brood resist. Some try to block the slavemakers' entrance; others pick up their prized brood and try to flee. The slavemakers are burly fighters, strongly built and equipped with deadly sabers. Each rebel is met and subdued by a single slavemaker who presents a simple choice: drop the brood, get out of the way, and flee with your life, or die. Many choose death. The slavemaker lifts her sickle mandibles to the head of the resisting worker, and the points of the mandibles are placed on either side of the brain cavity, poised for just an instant. Faced with such a choice, the worker holding a pupa may finally respond by dropping her sister. But if there is any hesitation, the points of the sickles meet through the resistor's brain, and the defender falls to the ground, dead. Stepping over the hapless defender, the slavemaker then grasps the pupal slave in her mandibles. Returning to the nest at a leisurely pace, she carries the initiate back to begin a new life of slavery—to be cared for and then duly initiated by members of her own species, fellow slaves. Of course, some pupae do not survive: They may be injured by the sharp mandibles of the amazons or may be eaten by slaves or fed to the amazons if there is a slave surplus.

Colony founding by the amazons occurs when mated amazon queens enter pure colonies of the slave ant species that are still independent. The amazon queen gets in by submissive posturing and gradually getting the workers to adopt her, probably with the use of appeasement substances. A week or so later the slavemaker queen approaches the resident queen and murders her in the typical amazon manner—piercing the head between her sharp-pointed mandibles. If the slaves feel any aggression toward their new queen, the murder of the old queen eliminates it.

For several years in his mid-teens, Hölldobler kept detailed notes, thinking that he had uncovered this remarkable behavior. Later he came across most of his "findings" in the early work of two Swiss entomologists from the previous century, Pierre Huber and Auguste Forel. Wheeler, too, was intrigued by the amazons, "one of the most beautiful ants," and wrote about them at poetic length in his 1910 volume *Ants.* But though beautiful in appearance, the amazons were truly degenerate. The only thing amazon ants can do well, as Wheeler summed it up, is fight.

The daily slave raids are conducted by sizable raiding parties on a scale surpassed only by army ants. In Forel's classic study of the Swiss amazons and their slaves, published in 1874, he found that a single amazon colony made forty-four raids in thirty-three days, retrieving an estimated forty thousand pupae and larvae. Of the forty-four raids, only seven came to nothing, nine gained a few pupae, and twenty-eight were complete triumphs. The slaves either move their colony or fight to the last female, the queen. Sometimes they pursue the amazons and try to get their brood back. Forel saw them succeed once in defeating the slavemakers, but it must be rare. Many other myrmecologists have since praised the amazon raids in glowing terms for their brilliant military precision and speed. As entomologist Caryl P. Haskins wrote, "The entire expedition is undertaken in concert, as though by word of command, and is carried through with a dispatch and definiteness of aim which almost compels one to believe in the existence of commands."

The amazon military communications system has since been investigated in the closely related amazon ant of Arizona. American Museum of Natural History entomologist Howard Topoff, who as a student worked with Schneirla on army ants, showed how the amazon scout locates new colonies and incites warring parties to go after them. It turns out that the slavemaker scout uses optical cues—the position of the sun and polarized light patterns in the sky in relation to bushes and rocks—to find and remember the location of prospective colonies to be raided. After locating the slave colony, the scout actually leads the way back out to the site. Only when the scout is trying to get the raid established does she lay a chemical recruitment trail for her nestmates to follow. All the raiding amazons contribute to this chemical trail, which helps to make a strong and decisive raid. After witnessing a few successful raids, Topoff moved some of the visual cues—led the ants astray—to prove these cues were crucial for navigation.

Topoff also found that the slave-raiding amazon queens produce special "propaganda substances" from their Dufour's glands. These substances are sprayed at and all around the workers defending their nest, and it allows the amazon queen to go into the nest and take slaves with comparative ease. Wilson, who with Fred Regnier of Purdue University was the first to identify this phenomenon in other slavemaking ants, mainly workers, found that such ants have greatly enlarged Dufour's glands full of acetates. The acetates are heavy and evaporate slowly, which means they exert influence over a long period of time. The "propaganda" works because the acetates mimic the chemical that is normally used by the slave colony for its alarm pheromone, but in such large quantities that the defending workers fly into a panic. In the confusion, the defenders fight and even kill their own nestmates.

In the five amazon ant species found variously in Europe, Russia, Japan, and North America, the details of slavemaking are similar. The amazon slavemaker ants, acccording to Hölldobler and Wilson in *The Ants,* are, depending on how you look at it, the "pinnacle . . . or nadir . . . of the slave-holding way of life." Other ant species sometimes take slaves, but they don't need them; that is, colonies of the species obtain profit or advantage from slaves, but they can thrive with or without them. Others take slaves by accident during raids on nearby colonies. These are considered the initial steps of the slavemaking habit. But the amazons and a few other ants have evolved toward dependence on their slaves. They find it difficult or impossible to live *without* them; they are physically and behaviorally degenerate.

In the early 1970s at Harvard, Wilson performed lab experiments to try to determine what would happen if these "obligate slavemakers" had their slaves artificially removed. Working with a tiny slavemaker called *Leptothorax duloticus,* which enslaves the tiny *Leptothorax curvispinosus,* Wilson first noted that the slavemakers depended on the slaves for food, which was gathered and regurgitated by their slaves. Similarly, just as in the amazon ants, the tiny slavemakers left all the brood nursing tasks to the slaves.

Next Wilson took the slaves away and watched. "Entire behaviors appeared for the first time," wrote Wilson. The workers soon began to move toward the brood, attending them and feeding them assorted broken eggs and pupal skins. They were solid materials only, but it was a start. The slavemakers allowed their brood to sit in partly shed skins for hours at a time, however—something the slaves would never have permitted. Meantime, some slavemakers began aimlessly carrying pieces of nest around in their mandibles, while still others began to feed by themselves on honey for the first time. Wilson noted that they took about ten times longer than the slaves to drink the same quantity and that the slavemakers never did retrieve solid food. Despite the changes in their behavior, the slaveless colony soon began to deteriorate. Wilson decided it was time to reintroduce the slaves. The result was instant: The slaves replaced the inept slavemakers in the brood areas and took over all the other jobs just as before. The slavemakers had their first good meal in weeks.

About the same time, Wilson got some startling glimpses into life in the slavemaker colony. He watched a slave approach the queen slavemaker and snap at her head and thorax, whereupon the slave worker laid an egg and placed it in one of the egg piles. Such insolence is probably largely symbolic. Yet unfertilized eggs can become males, which means that a slave might be able to get some of its genes into the next generation. Wilson saw this happen again later with another worker, with the same result. When a third worker tried it, however, the queen seized the egg and then fought over it with the worker. Each holding one end, they pulled it back and forth until it ruptured, whereupon the queen ate it. (In other ant species, two or more ant queens vying for dominant reproductive rights in a colony often pull out their rivals' eggs from the pile and eat them.) Another time, Wilson noticed a slave biting the queen on her right hind leg as she walked away from the egg pile. For twenty minutes the slave tried to drag her backward and sting her. Periodically, the slave made stridulation sounds. "All of these actions," Hölldobler and Wilson state in *The Ants,* "are typical of . . . workers engaged in fighting alien ants." They are aggressive actions.

Other slave ants have also been known to revolt at times. One slave species often bit its mistresses and dragged them out of their own nests. A few of the slavemaker workers lost parts of their legs in the attacks. Oddly, the same slaves who had attacked one mistress would groom and feed others, and slavemakers attacked by one slave would be cared for by other slaves. There was never any generalized revolt of the slaves. And never at any time did a slavemaker retaliate or precipitate an attack by going after one of the slaves. Other researchers have noted that the *Formica* slaves sometimes refuse to feed their amazon masters. The amazons mostly just stand around and do nothing—as usual.

As Wilson showed in his lab experiments, the degenerate slavemakers cannot survive on their own. Why don't the slaves simply take over? It would seem straightforward, but this longtime parasitic relationship is governed by fixed behavioral responses based on chemical and tactical cues. In the long and continuing saga of the ants there will always be slaves and slavemakers. If some species go extinct, new parasitic relationships will no doubt arise between other ant species. There are a number of ant species today that are seen as candidates for future slavery.

How did slavery originate in ants? It is a special kind of social parasitism—the coexistence in the same nest of two ant species, one of which is a parasite dependent on the host. When the social parasite is a warlike ant that raids other nests, captures the brood, and brings them back to rear them in its own nest, the result is slavery. The slavemakers are, in effect, warlike, flamboyant social parasites.

Darwin read Pierre Huber's original descriptions of the slavemaker/amazon ants published in 1810 and was fascinated. He developed the first hypothesis to explain the origin of ant slavery, which he included in *On the Origin of Species* in 1859: Slavemaking ants started when warring ants began habitually raiding another closely related species in order to obtain its pupae for food. In time, some of the pupae emerged as workers in the foreign colony's nest before they could be eaten. They were accepted as nestmates. The expanded workforce benefited the superorganism—enough to encourage natural selection to promote the tendency for colonies to start raiding specifically to obtain slaves.

Father Wasmann, among others, rejected Darwin's suggestions as a matter of course. The real story, however, seems to lead back along the route that Darwin forged. Summing up current thinking on the subject, Hölldobler and Wilson recently outlined three hypotheses for the origin of slavery. Besides Darwin's brood predation hypothesis, the other two are that territorial raids between colonies of initially the same and later closely related species led to the robbing of rival nests, the taking of brood, and the occasional hatching of this brood before eventual enslavement; and that ants that often transport brood within a multiple-nesting colony might begin to do so with other closely related species, again leading to slavery.

According to Hölldobler and Wilson, territorial behavior is probably the key factor or spark that leads to slavery. In their most likely scenario, the territorial raids "combined with a strong propensity to transport brood lead to the regular retrieval of alien brood back to the raiders' nest; the raiders destroy and eat most of the captives, but a few survive to join the colony as slave workers."

The slavemaking habit has evolved at least eight separate times in the ants: twice in the formicine and six in the myrmicine ants. But the overall phenomenon of social parasitism, of which slavemaking is only the most notorious aspect, is more widespread: At least two hundred species are known, and more are being discovered every year. The preeminent ant social parasite hunter and German entomologist Alfred Buschinger and his various colleagues have identified several predisposing features of colony structure and the environment to explain why certain ants may turn to social parasitism and slavery.

Typically, such ants live in colonies with multiple queens and have a practice of readopting newly mated queens. At least some of the multiple nests they occupy periodically have no resident queen. Both of these structural factors enable colonies to be more easily invaded. In addition, the colonies may not be born with the ability to recognize their own species' odor but must learn it. This means they can be accepted and learn another colony's odor. They also live in dense populations.

Usually, the likely social parasites and their hosts live in the deserts or cooler temperate areas. Scientists have long puzzled over this, and until recently many considered that the lack of information about tropical social parasites was simply because little work had been done in the tropics, but as this gap has started to narrow due to the work of Wilson, Brown, Hölldobler, Longino, and others, the cooler temperate zone and deserts are now seen more and more as a prerequisite. It may be that cool or dry conditions are crucial to dulling the senses of the ants and reducing their aggression and intolerance of foreign species in their own nests. When a myrmecologist working experimentally in the lab wants to introduce foreign workers to a colony queen, one technique is to chill the queen beforehand. She becomes more tolerant, and by the time she warms up, the workers have started acquiring the colony odor. But the paucity of tropical social parasites may also be that densely populated colonies occur less often in the diverse tropics. In any case, as more and more research focuses on rare tropical fauna, there will be more social parasites found there—though it is unlikely to ever rival Switzerland, where a full one-third of the 110 ant species are parasitic.

A final crucial factor in one species becoming a social parasite or slavemaker of another is that they be closely related. This is true of all permanent social parasite relationships between ant species. This was first pointed out in 1909 by Italian myrmecologist Carlo Emery, and it is now referred to as "Emery's rule." It helps explain how two closely related species can come to live together in the same nest.

Emery suggested that the slavemakers as well as other permanent social parasites actually evolved from the slave or host species. This idea seemed preposterous to Wheeler and others at the time, but Wilson explained in 1971 that it may well be true in the way that we now understand geographic speciation. A population of a single species becomes geographically separated, and the two evolve in isolation into two species. At a later date they meet as two different but closely related species and compete fiercely for food and space; one proceeds to fight the other, take larvae, and sometimes, if the larvae survives to work in the alien colony, to become slavemakers. If this relationship persists, then the two species may coevolve, with one becoming more specialized as slavemakers and the other as slaves. Because the slavemakers have more derived characteristics than the slaves, it is more likely that the slavemakers have evolved from the slaves than the other way around. This leads to the bizarre but strong possibility that the slave has actually given rise to her own mistress.

HÖLLDOBLER NEVER GOT to publish his boyhood notes on amazon ants in Germany, but he did go on to make his name with other slavemakers. His study of one slavemaker ant in particular, the honeypot ants of the American Southwest deserts, really made his mark in myrmecology. These are true slavemakers because they actually enslave members of other colonies of their own species. The slavery they practice is only a small part of some of the most complex strategic wars yet discovered in ants. These ants are skilled practitioners in the art of measuring the enemy. During the Gulf War, Saddam Hussein could have taken a few lessons from the desert strategies of these ants.

In the early 1970s, soon after Hölldobler moved to the United States to become a professor at Harvard, he began to make regular field trips to Arizona. There he met the honeypot ants, and he soon wanted to learn everything he could about them.

Like the weaver ants, leafcutters, army ants, and imported fire ants, the honeypot ants *(Myrmecocystus)*, Hölldobler knew, had already achieved some notoriety among entomologists for the bizarre appearance and the behavior of certain workers who literally become "honeypots," hanging engorged from the roofs of their underground nest. Their fame dates from 1882 when the Reverend Henry C. McCook published his book *The Honey Ants of the Gar-*

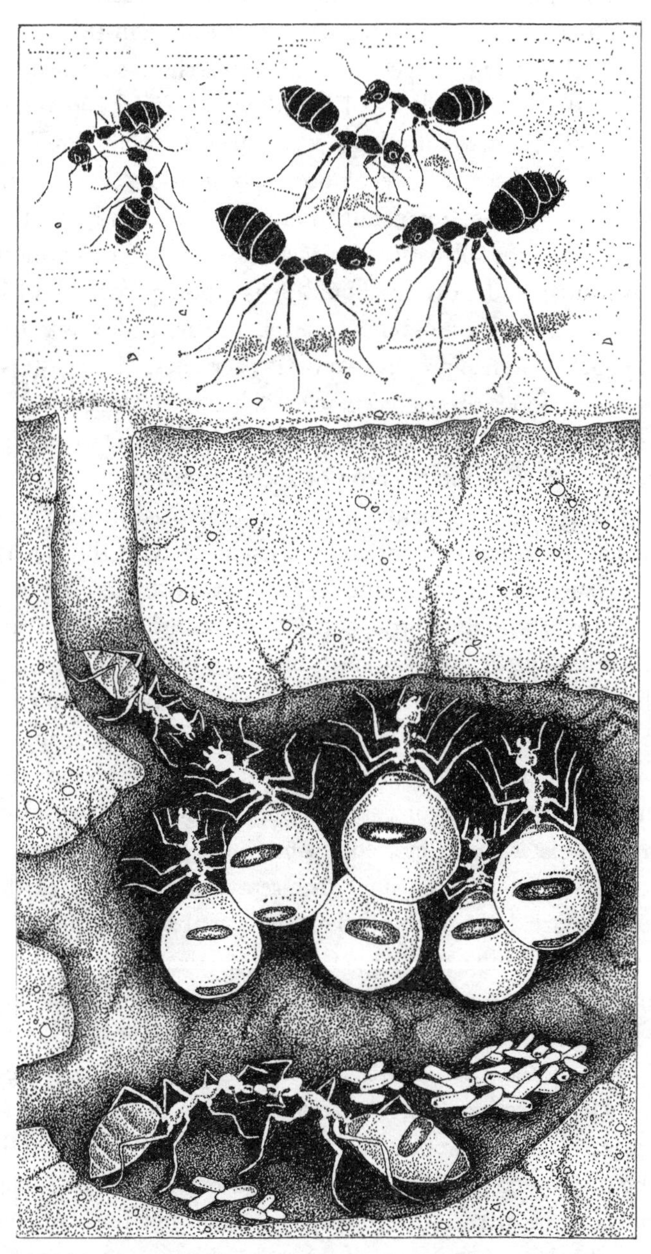

The Honeypot Ants. On the desert floor, two colonies of honeypot ants, Myrmecocystus mimicus, *square off to conduct a nonviolent minitournament. Sometimes they stand on tiny pebbles to gain height advantage. Down below, in the nest of one of the colonies, a foraging ant regurgitates the liquid from her crop. The liquid is stored in living "honeypots"—ants who hang from the roof of their underground bunkers.*

den of the Gods and the Occident—though the first description of them appeared in an obscure Mexican journal fifty years earlier, in 1832.

In *Ants,* Wheeler brought together all the research on so-called honey ants. He noted that the desert Indians of Northwest Mexico and the Southwest United States were familiar with them as an esteemed food item. The object of their desire was the special caste of honeypot workers with their enlarged golden gasters full of sweet, juicy, nutritious liquid. At first it was thought that the ants manufacture the substance, but McCook, refuting that notion, revealed how the ants obtain and store the liquid. It is a potent cocktail of honeydew obtained from various homopteran insects, nectar, and water from desert plants, and liquefied animal protein—typically termites. Their gasters, in effect, function as colony food storage receptacles. When the ordinary workers return to the underground nests, they approach the honeypot caste to regurgitate their catch, which is then stored in the crop or social stomach.

This practice is found in many ants and other social insects; it is the way food is transported around an ant colony, particularly to those who can't go out and hunt, such as the queen and larvae. But most ants have gasters with crops that will expand only a little. Wheeler identified ants of various worldwide species with enlarged gasters, most of which live in the desert, and called them all honey ants. But Wheeler focused on the North American desert honeypots (various *Myrmecocystus* species). Far beyond the capability of other ants, they absorb more and more of the fluid from literally hundreds of workers until the expanding gaster of a single honeypot worker swells to the size of a chickpea. The gaster varies in color from golden yellow to amber, but it is possible to see brownish black lines that represent the color of the original gaster before it started expanding. When she can absorb no more, she climbs up on the chamber ceiling as if to get out of the way and hangs there along with dozens of other honeypots. In this way the food for the colony can be stored for months as a hedge against uncertain food supplies and lean times in the desert; in effect, the ants are airtight sterile containers. Each honeypot caste worker hangs by her claws from the rough vaulting of the roof, side by side, hundreds in each large chamber. Occasionally they move a little, but if one falls off, she needs help, as Wheeler put it, "to regain the pendant position." These honeypots are fiercely defended, of course, but they are also able to defend themselves to some extent, if necessary, though they never leave the nest. They have mandibles. A storage jar that bites—yet another brilliant invention of the ant world.

Hölldobler brought back a few colonies of honeypot ants to his lab at Harvard. Little did he know that one of these colonies, with its queen, would stay with him for more than eleven years at Harvard—longer even than the most attached graduate student. Once they were settled in his lab, he found that if he gave the workers a roof to hang from, some would turn into honeypots. These were usually the largest workers who probably began serving this role as callow adults when their bodies were still soft and elastic. But he really wanted to find out more about their slavemaking and warlike behavior. Down the hall from where Wilson was taking slaves away from his tiny *Leptothorax* slavemakers, Hölldobler began staging queen competition and fights. Over the next decade the lab wing of the Museum of Comparative Zoology would see some odd wars based on endless posturing. But arguably the more interesting wars—as well as strange ant tournaments—were happening out in the field.

Walking through Arizona's mesquite-acacia community after the first of the summer rains had softened the hard, dry desert, Hölldobler found lots of honeypot colonies new and old. He confirmed that the honeypots lived in much deeper nests than Wheeler had ever envisaged. Working in Arizona in the 1950s, Robert H. Crandall, assisted by local gravediggers, dug sixteen feet down until he found the honeypot queen at the bottom of the nest in a small, heavily guarded chamber. He counted fifteen hundred of the honeypot caste workers, although Hölldobler believes that two thousand honeypots may be a good figure for a typical mature colony of fifteen thousand workers and one queen.

The story of the honeypots, for Hölldobler, was just starting. In 1974 he was lying on his stomach in the Arizona desert southeast of Tucson when he noticed the honeypot ants congregating on the desert floor and performing strange movements. It was mostly the larger workers of the colony—those that had not become honeypots—but they appeared even larger than usual. They seemed to be walking on stilt legs, as if to make themselves appear larger. Some climbed on pebbles or bits of raised ground to get even more height. Then they turned to face each other, one on one, before slowly turning to one side. They began sparring. After this initial face-off, the pair would slowly circle each other, all the time kicking out at the other with their legs and drumming with their antennae on the other's body. Then came some pushing and shoving. Eventually, one or the other would give up and retreat, or walk away to find another ant to spar with. Even with several hundred ants facing off, there were no mandibular knives pulled, no chemical mace sprayed, no violence. And when two ants from the same colony met, they simply checked out each other's pheromones by sweeping their antennae back and forth. When they recognized each other, they bobbed up and down as if to say "Hi!," then moved on.

The ants almost seemed to be playing, Hölldobler thought, even though he knew that ants do not play. He noticed another part of the colony going off on a termite hunting expedition while the first contingent kept threatening the other colony. He began to get the picture. One part of the colony was keeping the competition busy while it monopolized the local food resource. But just what were they doing with this nonviolent sparring? Why bluff the competition when you can fight and kill them? Or perhaps the question should be rephrased: Why fight the competition and die in some numbers when you can bluff them and win? Sometimes it went on for hours, even days. What did it all mean?

Hölldobler resolved to find out. As ritualistic as it looked, he never really thought about ants being capable of advanced rituals. For weeks, baked by the heat day after day, he lay on his stomach watching the ants and taking movie film of the ants' activities. Using his strong visual sense and eye for detail, he became a macro-cinematographer. Sand blew into the cameras and into his face, but still he persisted. Back at Harvard, he played and replayed the films, first in real time, then over and over again in slow motion. He mapped the movements of the ants in detail until a scenario emerged.

This was indeed some bizarre ritual. As the ants, seeming to be on stilts, faced each other in frame after frame, the purpose of their ritualized behavior was revealed: More than just big bluffing contests, these were tournaments in which the ants of neighboring colonies assessed one another's strengths and weaknesses, and the strongest colonies would, in effect, stop the others' expansion or force them to retreat into smaller and less favorable territories. The territories are not fixed but are "spatiotemporal"—temporary territories chosen depending on where and when prey is present. Other ant species aggressively defend permanent territories, but life in the desert dictates resourcefulness and flexibility. In the same way that humans survive in the desert through flexibility—moving as needed, sometimes over large areas, to secure food and water supplies—many animals and even insects do the same.

As Hölldobler studied the development of the tournaments, how they begin and how they form gradually into tournaments that may lead to war, he discovered that each colony posts guards along the frontiers and boundary areas between its territory and those of others. Fewer than a dozen guards stand high in their stilt postures for hours. When the bordering colonies dispatch their own guards to the area, minitournaments sometimes occur that last up to several weeks. Only when one colony greatly increases the number of guards does the other colony send for fresh replacements, and then a full-scale tournament may be enacted. Rarely did violence occur during a tournament. They were basically nonviolent affairs. Yet often there were clear winners as one colony backed off.

But Hölldobler had a nagging question. The tournaments made sense only if the ants could correctly assess their opponent's strength. Can a colony of ants determine whether another colony is stronger or weaker than itself?

Hölldobler started counting ant heads and filling out scorecards for dozens and dozens of tournaments. With his film footage of the contests, he was able to achieve a high degree of accuracy. He found that the ants did correctly assess the other's strengths and weaknesses but wondered how they could manage it without an aerial and underground view of the proceedings—without the sort of intelligence human armies get through aerial photos from spy satellites or through infiltration. How was it possible?

Showing the films and his scorecards to one of Wilson's collaborators, Charles Lumsden, a physicist and theoretical biologist from the University of Toronto, Hölldobler came up with three main ways the ants might assess each other's strength:

1. They could count heads as they face one opponent after another. If there were a substantial discrepancy, with one army finding far fewer sparring partners, this could be the signal to become more aggressive and, for the weaker colony, to beat a hasty retreat.

2. The ants of each colony could "poll" each other. If the ants of the foreign colony included a lot of majors, the largest workers, it is likely a large colony. Majors are produced in great number by the mature colonies.

3. The ants could try to judge how long it took to find an opponent to face off against. If it took a long time, that means there are fewer ants in the opposing colony, and it's probably weaker. A short waiting time would mean there are numerous ants in the other colony, and it's strong—time to retreat.

After many more days staring at his desert insect movies, Hölldobler determined that to some extent "all three measurement techniques are used." The small immature colonies "probably concentrate on the caste polling method [which] gives them the quickest indication of whether the opposing force is relatively large, so they can make a rapid and sensible retreat."

A single tournament may last for days, with breaks each night when these ants are inactive. Short tournaments are usually due to lopsided contests in which the weaker colony retreats and the strong realizes that the time has come for some real old-fashioned combat. The attacker may have ten times as many warriors as the colony being attacked; therefore, the aggressor knows her enemy so well and is so sure of victory that the resulting war is entirely one-sided. The victors slice up their enemies and race deep into the nest. They carry off the brood for slaves. Then they enter the golden honeypot caverns and dislodge the big honeypots from the roof and drag them away. These are the real spoils of war. Finally, as they reach the deepest heart of the nest, they dismember and kill the hapless queen—all with minimal loss of their own colony's life. The Allied forces in the Gulf War had "smart bombs"; the honeypots wage "smart wars." As Sun-tzu, the Chinese general and military strategist, wrote in *The Art of War* more than twenty-three centuries ago: "Know your enemy, know yourself, know the ground.... Your victory will then be complete." By all accounts and measurements, the tournament honeypot ants of the American desert are a formidable adversary.

Survival in the desert requires not only fancy footwork but may rely on bending the "rules" of "normal" behavior. The honeypot ants also turn out to be one of those ant species that resort to robbery to increase their food supplies. Hölldobler watched his honeypots in the Arizona desert as they, singly and in small robber bands, routinely stopped harvester ants, *Pogonomyrmex,* as they worked hard to carry seeds and other plant materials, and sometimes fresh termites, back to their own nests. The honeypots seem to lie in wait near the entrance to the harvester nests or at other spots where they pass. Like customs inspectors, the honeypots let harvester ants with plant materials pass, but any carrying the precious termites have their loads considerably lightened. If the victim decides to protest the confiscation, sometimes trying to bite the robbers, the honeypots simply scamper away and wait for the next harvester. Hölldobler calls such thievery a specialized form of aggression.

Even before Wilson collaborated with Hölldobler on their classic ant book, he remarked that the younger man's work on the honeypot ants and their tournaments "is, in my opinion, right up there with the waggle dance of the honeybee in the quality of the research and the magnitude of the phenomenon."

The fast, agile honeypot ants are slavemakers and robbers. They stage elaborate ritual tournaments as part of a brilliant strategy for war and domination. They distract and deceive other colonies—they're cheaters. They cleverly store food in sterile living honeypots—the ultimate use of the social crop of ants—as a hedge against hard times in the desert. Robbery, cheating, slavemaking, food hoarding, jousting, fighting, killing, and all-out warring—they stop at nothing. But even the most resourceful ants in the world get at least a little comeuppance.

In the interstices of the large, successful honeypot ant colonies lives another fast-moving desert ant, a species called *Forelius pruinosus.* A worker is less than one-tenth the size of the honeypot worker, yet these ants manage to get their share of the precious termites and other food in the desert. They gather around the food and exude clouds of noxious chemicals to intimidate the honeypots. They often steal the honeypot ants' food right from under their mandibles. At times the tear gas–like clouds are so thick when the ants mass in great numbers around the entrances to the honeypot ant nests that the honeypot ants are forced to remain underground while the tiny ants scavenge food all around. This greatly extends the little ants' own hunting grounds.

Another nemesis of the honeypot ants is *Conomyrma bicolor.* These tiny ants also gather in large numbers and use chemical weapons to immobilize the big honeypots while they hunt at leisure. But these ants have an added technique for keeping the honeypots down underground and occupied, or at least distracted. The workers pick up pebbles in their mandibles and drop them down the entrance holes. This strategy reduces the time the honeypot ants spend foraging out of the nest. As Hölldobler and Wilson point out, this can be classified as "tool use"—a rare example not only for insects but for all animals. It is a tribute to these ants that they have evolved so many strategies not only to survive but to flourish in response to the difficult environment of the desert.

THE FINAL WAR for control of the aztec world, located in the maturing cecropia tree not far from the new lab at La Selva and above the massive leafcutter colony, is almost at hand. The tall, sun-seeking cecropia, pushing fifty feet into the air and almost a canopy tree after only six years' growth, bears little resemblance to the scrub tree that represents the humble, peaceful beginnings of all cecropia trees. Cecropias are sometimes called trumpet trees. The angle of their branching pattern, arching upward toward the sky, resembles a trumpet. The tree's trumpet posture is simply enthusiasm to reach the sun. Cecropias *love* sun. Normally, they won't even bother sprouting until burning, cutting, or natural tree falls create a sunlit opening in the forest. Thereafter, they will do anything to reach sunlight. In fact, they sacrifice structural strength in the trunk, erecting a simple hollow frame and manufacturing leaves with few defensive chemicals in an effort to focus on growth to the canopy. Even when they do reach the canopy, cecropias live fast and often die young.

This cecropia has a double trumpet due to the fork near the base of the trunk. Cecropias such as this one that have become symbiotically involved with aztec ants—not all of them do—eventually turn antisocial toward all.

Peaceful at first, the aztecs metamorphose into adolescent warriors, and by the time the superorganism is maturing, they are professional soldiers ready to fight anything that so much as brushes against the tree or is caught wandering inside a radius of up to sixty feet. Tropical ecologist Daniel Janzen described these ants as the chemical response of the tree to its predators. To these ants, the tree is their entire world and the battleground on which they fight for world dominion. But not only do they fight other species of ants and small mammals, frogs, lizards, and humans, they also fight other colonies of their own species—and reserve for them a special ferocity. In this all-out war they take no prisoners or slaves. There is neither compromise nor room for peaceful coexistence. It is ultimate victory or ultimate defeat.

This particular aztec-cecropia is inhabited by two colonies, one old, massive, sprawling colony in the western fork of the tree, and the other younger, fiercer colony living just beneath the canopy top in the eastern fork of the tree. It had started with one colony and soon grew to eleven, plus a few other colonies of other ant species. But each in turn had been bumped off by a colony more aggressive and powerful than itself.

In recent months there have been skirmishes between the last two remaining aztec superorganisms. The build-up to a final war has begun. The story of how an aztec superorganism conquers a cecropia, selfishly excluding all other aztecs until it reigns supreme, is a multiyear saga filled with forced evictions, numerous treacheries, blatant violence, bullying, occasional kidnappings, and pitched battles—all directed toward other colonies of their own species. As with the tournament honeypot ants, this is war even in the strict terms humans use. Unlike the honeypots, this is the dirty, seamy side of war—all hyper-aggression, no finesse.

Following is a typical chronicle of the life of a cecropia tree and the power struggles of the developing aztec superorganisms in their quest for ultimate victory.

Minus two years and counting. In a mass production of some nine hundred thousand seeds, a thirty-year-old female cecropia tree at La Selva dispatches its genes into the next generation. It is not the only or even the best seed production in the comparatively long life of this tree, which has had frequent reproductive episodes in the previous few years. But the strong and sudden winds that knock down the tree ensure that this is the last and most effective reproductive attempt. The old aztec colony inside the fallen cecropia dwindles and dies, unable to escape as the queen's chambers are blocked by debris after the fall. But the cecropia does better; its seeds are blown far and wide. The seeds are variously eaten, carried away, or covered with leaves. One seed that lands at the edge of the closed canopy forest has a good chance.

Birth. After heavy rains and a couple of tree falls, the sun bakes the earth for several days. Just beneath the surface, the cecropia seed, dormant for two years, sprouts. It pushes up through the fine leaf litter and it grows three feet through the first summer rainy season. By the end of the year it is eight feet tall, with a pronounced fork that splits the tree at about four feet above the ground.

Age one. As the tree tops twelve feet, the fork becomes more defined; one branch extends up at a slightly easterly angle toward the center of the light gap, and the other fork moves almost straight up, slightly to the west. Near the top of each fork, in the area that receives maximum sunshine, the production of Müllerian bodies, the whitish yellow corpuscles at the base of the leaves, begins to attract attention. First, various birds come to check them out, nibbling at the sweet nutritious food bodies. Woodpeckers come to pick up unsuspecting insects but go away empty-mouthed. Then an aztec queen notices the tree and climbs aboard. This is precisely what the cecropia had in mind. The first queen to move in is a veteran queen with a small workforce and a few head of mealybug cattle—a colony that recently lost its residence in a mature cecropia. They were lucky to escape alive when the big flood moved through the area. They move into the western fork of the tree. The queen notches a hole through a weak part of the trunk underneath a branch and slips in. She and her workers use some leaves to chew a carton paste to remodel the interior of the tree. For the moment it is just a shell, but they soon turn it into a home with brood nursery and separate queen and workers' quarters.

A few days later a newly mated aztec queen climbs aboard the tree on its eastern fork and finds a cozy spot near the top of the branch. Unlike the older queen who has workers and a few mealybug cattle, the new queen must do all the work herself and build up her resources. The tree wants to have these protectors on board. Besides the daily lines of foragers that pluck the Müllerian bodies from the leaves and bring them back to the colonies, an increasing number of aztecs patrol the leaf, stem, and trunk surfaces, nibbling any vines or other plants that would try to gain a foothold on the cecropia. In this way the aztecs ensure that the cecropia has its best chance to reach the canopy and that no unnecessary weight robs the tree of its vitality.

Age two. Eleven colonies have taken up residence in the rapidly growing cecropia, about evenly divided between the western and eastern forks. Eleven superorganisms. Eleven animals with their own individual interests paramount. Initially, none is aware of the others. Each settles on the tree as if it will be able to stay or become the sole resident. There are no apparent problems, no conflicts. The tree continues to grow fast, eight inches or more a month, putting on enough new growth to guarantee the steady addition of gallery space. Leaf production is high enough to support all the various colonies and their needs for Müllerian bodies, as well as occasional harvesting by the leafcutters of shoots near the base of the trunk. However, there is growing inequity in the mealybug cattle kept by the ants. Some have several or more; others have none. Those that have several are more aggressive. The sweet honeydew the mealybug cattle produce is lapped up by the ants. The more they get, the more excited they are to get more. The honeydew gives these ants their source of quick energy that keeps them so highly charged. The cost is a lot of time spent fighting instead of working and often a short life span. But possessing the fuel for war is one of the most important indicators of a colony's eventual success.

In eight months the five largest, fiercest colonies kill or drive off the six weakest, one by one. There are some losses among all the victorious colonies, but they soon build up their fighting forces again, preparing for the next round of battles. They fight by day, and all the work on the tree, collecting Müllerian bodies and patrolling outside on the trunk, occurs by day, though often with a period of inactivity in early afternoon. At night the ants stay inside their galleries and things are quiet. Initially the day battles are more frequent, but as fewer and fewer colonies occupy the tree, the stakes get higher and the last few colonies try to wait for the opportune moment for attack. Months, even years, sometimes pass between battles, but there are always more.

Besides the aztec ants, a colony of cork-headed ants *(Zacryptocerus)* has settled on one open stretch of this tree, two-thirds of the way up the eastern trunk. Whenever the aztecs approach, the cork-headed workers slip inside the tree and plug up the hole with their flat heads. Each time the aztecs inspect the area, they can find nothing although the place clearly smells of other ants. But the eastern fork colony of aztecs eventually drill through the inner walls of the tree just above the cork-headed ant colony, penetrating it to the heart. It's all over in a few minutes. They corner the queen. The cork-headed ants try to defend her, knocking against the aztecs using their large angular heads like shields, but the vicious aztecs bite at the ants, pushing them out of the nest and driving them off the tree. Finally, they concentrate on the cork-headed queen, surrounding her and spread-eagling her. In seconds they execute her, chopping off her head.

Age four and a half. At thirty-five feet high, the cecropia, thanks to the aztecs patrolling its surface, keeps pace with the other plants around it in its quest for the sun. Strangler figs have no chance to gain a foothold on this tree. The aztecs are as keen as the cecropia to make sure it reaches and maintains a spot in the canopy. The sun's energy plays a crucial role in making more and more of the sweet Müllerian bodies that the aztecs need to supplement their honeydew mealybug rations.

Late in the year, with only three aztec colonies left, the most spectacular battle to date sees thousands of casualties of the old aztec superorganism situated midway up the western fork of the tree. More than 150 majors are killed; few now survive. The old western colony located high in the tree is the aggressor. In the final hours before the defeated queen risks leaving her chambers and fleeing, much of the brood is abandoned. The queen emerges, is surrounded by five enemy aztecs, and bitten and sprayed and then pushed from the tree. She tumbles and falls eighteen feet to the leaf litter. At first she is only stunned, but even with no movement a patrolling leafcutter soldier from the nearby colony smells her, releases her extra alarm pheromone, and within seconds fourteen leafcutter soldiers surround the fallen queen and kill her easily. She doesn't resist. They leave her corpse, but half a day later, as the necrophoric smells begin to be released, they carry it to one of the leafcutter waste piles behind the nest. The third-to-last aztec colony has fallen and will never regain its territory.

Age five. As the cecropia approaches the canopy, the two remaining aztec colonies prepare for the mother of all ant battles. No matter how tall the cecropia grows, it still is not big enough for more than one aztec colony. For weeks there are the warning drumbeats of war, skirmishes here and there.

The skirmish that leads to the last all-out conflagration occurs at dawn when a young aztec warrior from the western colony slips and falls while carrying some mealybug cattle. She falls onto a branch ten feet below that belongs to the eastern colony. She is only a few feet inside enemy lines, but she is clearly in danger. She stands up, picks up the shaken mealybug in her mandibles, and starts to walk down the tree, attempting to cross the DMZ on foot and go back up to the western fork of the tree to safety. Only a few feet from the border she meets seven eastern aztec warriors on patrol. She is spread-eagled and restrained. She drops her mealybug, which is quickly picked up and taken from her by one of the warriors. Putting her head back, she sprays a massive blast of alarm pheromone into the air, and then the blades of her enemy are upon her, dicing her into small pieces. But as the news of her danger—the big blast of alarm pheromone—drifts across the DMZ, the western colony springs into action. The final war for control of the entire cecropia tree is now under way.

Hundreds of soldiers, their mandibles open, their sabers drawn, charge down the western fork of the tree. They round the base and then head up the eastern fork, killing, maiming, dispatching first the half-dozen or so soldiers that killed their nestmate and then the dozens that stand behind them. The attacks on so many ants releases ever more pheromones into the air, summoning more and more eastern soldiers to the fray. Ants from both sides charge up the trunks and enter the galleries to call to the soldiers in their bunkers to come out and fight. The galleries of the tree and the battlefield of the trunk become suffused with the odor of aztec war, the various cyclic ketones with their pungent smells. The concentrated air from the emissions of hundreds of soldiers, as in a major battle, can be truly repugnant.

By mid-morning the battle escalates, and now there are thousands of soldiers engaged in fighting. At first, most of the fighting occurs on the eastern fork of the tree. But with the east's advantage of height, fighting the west as it advances up the eastern fork, and perhaps a measure of home field advantage, the effect of the west's initial surprise attack begins to weaken. The west is the older colony in this case, but that is not necessarily an advantage or disadvantage.

Meantime, behind the front lines, many workers in the east and west continue to work, collecting Müllerian bodies on the leaves in the canopy and bringing them down to the soldiers just behind the line. Even more important is the "milking" of the mealybugs and the carrying of fresh sweet nectar to the soldiers. It is about a forty-five-foot walk from the canopy and highest galleries in the western colony to the DMZ near the base. The eastern fork of the tree is about thirty-nine feet tall. The soldiers on the eastern fork are supplied a little quicker, especially since the fight has been mostly on their home ground, but they have fewer galleries than on the western fork.

With so many ants out in the open, their homes ripped apart and everywhere the smells of war, a woodpecker, an insectivore, has come to eat its fill. Most mammals will not put up with the aztecs, except for sloths and a few other herbivores that feed on cecropia leaves, but certain woodpeckers will go after the aztecs themselves. A woodpecker can catch and swallow them before they do any damage. Ants may not be the tastiest food in the jungle, but they are a ready source of protein. And the aztecs full of honeydew can even be sweet tasting.

Then two more woodpeckers circle the tree before swooping in. Right in the middle of the battlefield, the three woodpeckers peck at the tree, pulling off workers one by one, indiscriminately taking members of both ant colonies. But they are also pecking holes in the tree and helping themselves to ants still in their galleries. A woodpecker who manages to get a queen can easily swing the final battle for this tree in a single stroke. One woodpecker hammers into a gallery that adjoins the queen's quarters, opening up a rich source of fresh ant larvae and pupae. But no queen. Gorging itself, the woodpecker then flies off. All the birds are soon satiated or tired of their ant diet and depart as abruptly as they arrived, disappearing into the darkness of the closed canopy.

Although they are large predators, they have little effect on the overall course of the aztec wars or on the number of casualties.

By noon the western aztecs have taken nine galleries on the eastern fork, turning out the eggs, larvae, and even a few pupae, and killing the nurse-workers. Only a few pupae are carried to safety farther up the tree.

At about 3 P.M. the line of attacking western aztecs begins to thin. Their many casualties and the greater distance from their supplies is taking its toll. More than a hundred young, freshly sugar-watered, fierce eastern soldiers break through the line and charge down the tree, rounding the fork at the base and heading up the western fork to take the war to the enemy. They pass numerous of their own and the enemy's dead in the DMZ before meeting groups of western nurse-workers carrying brood. They are heading *toward* the enemy, not away. They seem to be attempting to move into the galleries vacated by the dead or retreating eastern colony following the first wave of the attack. Their effort to supplant the eastern colony is a little premature. The babies and their nurses are clearly in the firing line and are disposed of with ease, most merely pushed off the tree. The scavengers—provided the leafcutters below don't sweep everything up—will have a fine time tonight.

Moving up the western fork of the tree, the eastern soldiers advance with an assured ferocity as they invade each gallery, killing or forcing out the workers and turning out the brood. As new chemical messages are relayed around the western fork of the colony, a new reserve soldier force is recruited to attack the invaders. The west is still stronger than the east.

All these battles mean little, of course, unless the queen herself can be found and driven out with her guards and killed. The casualty of one queen is more important to the survival of a colony than a hundred thousand dead workers and brood. To be left with only one queen, a colony can theoretically survive. One queen is worth the entire game, although as a practical matter, without a few workers to assist in raising new broods, it is difficult to carry on.

In the final hours of the day, as the western soldiers begin to make a serious dent in the eastern colony's advance and its numbers, the eastern soldiers suddenly strike gold. Just below the canopy, they probe a rather large hole to find the queen and the queen's guard surrounding her. The air is ripe with defensive chemicals and, soon, death. The eastern soldiers, their mandibles snapping, bite the guards, squirting poison everywhere. When the air clears, the old western queen is left to defend herself. She stands up as if to show her size. She has nowhere to run. Surrounding her, six eastern aztecs spread-eagle her and perform their butchery. From her swollen gaster, as it is split open, the whitish eggs of thousands of unborn aztecs spill out onto the chamber floor.

She struggles, her face buried up to her eyes in her own unborn progeny. Her antennae swirl around and around the mass of eggs as if searching for some familiar smell, some sign that help is on the way. And then she is still. The eastern soldiers move on. They have done the most important part of their job. Now they must finish it. For even though they have killed the queen, there are still many more workers and soldiers ready to fight. In a short time, however, the colony will dissipate and become disorganized without the queen. The war is all but over now.

In the months after the last aztec soldier evacuates the western colony to spend her last days wandering aimlessly around the forest, everything is suddenly quiet on the cecropia. The eastern aztecs have won. As they occupy the galleries in the western fork of the tree, life becomes almost mundane. There is no one else to fight. Of course, there will still be insects that carelessly land on the tree and scientists who dare to touch. The aztecs are only too happy to dispatch them, and they may well expand their domain around the base of the tree, incurring battles with the leafcutters, but for the most part they are now a totalitarian society. The colony may live another five years or ten, maybe more. It depends on the health of the queen. She must avoid falling victim to a woodpecker attack. But most of all, it depends on the tree's health. In fact, the more the ants inhabit the tree's galleries, making the trunk more and more hollow, the more the tree is weakened structurally. The cecropia is a fast-growing tree that lives ten to fifteen years ordinarily, and occasionally up to thirty or thirty-five years—if it is not cut down or blown down first. For now, this cecropia is healthy, strong, and growing, with a dozen or more years left. The conquering workers lift their antennae as the tree itself raises its trumpeting branches to the sky as if to herald the eastern aztec's final victory.

THE GREAT ANT wars are as old as the social life of the ants—at least 100 million years old. Today, ant wars are still being waged with the same primitive ferocity by more derived ant species, employing their bizarrely adapted mandibles and their chemical warfare to rule the Earth. Other insects, birds, mammals, humans, all creatures of more recent origin, largely can only stand by and watch from the sidelines. A few conduct their own minor battles, of course, all of which are comparatively insignificant affairs with the exception of humans.

Most ant wars are little wars of checks and balances without much territorial gain or loss. Most ant species—species in the subfamilies of the ponerines, early myrmicines, and even the dorylines—are nonexpansionist. Some of the Dolichoderinae ants, such as the aztecs, will battle to the bitter end over a single cecropia tree. The aztecs may take a whole tree, but ten or fifteen years later, on average, when the tree falls and dies, they often die, too, or evacuate and try to find a new tree and start again. But the big, colonial fascist species —as myrmecologist Caryl P. Haskins dubbed them in his 1945 book *Of Ants and Men*—the little fire ants *(Wasmannia auropunctata)* and the Pharaoh's ants *(Monomorium pharaonis),* the imported fire ants in the American south *(Solenopsis invicta),* the Argentine ant *(Iridomyrmex humilis),* and *Pheidole megacephala,* all advanced myrmicines, and the formicines, are the true warmongers bent on world conquest.

Far from the aztec world and its victory in the cecropia tree, far away in space and time, some of these ant conquerors land on a beachhead. For three days they smelled land as they drifted closer and closer to the island. It turns out to be the island of Hispaniola, "discovered" by Columbus on his first voyage to the New World, albeit many centuries after the first Native Indians discovered it and millions of years after the first ants arrived. At 4 A.M., with no fanfare, the nearly waterlogged craft puts ashore just short of the high tide line on the south coast of the Dominican Republic, which shares Hispaniola with Haiti. The landing is a few miles east of the capital city of Santo Domingo, near the international airport. The ant colony is exhausted. Six workers never get up from their bunks. They are dead on arrival. The queen is alive but weak and, with little food, has stopped laying eggs. Within an hour the first worker leaves the raft to search for food. Five minutes later others depart, and within fifteen minutes the three strongest workers find a large water beetle turned over on its back. Smelling and prodding it, the ants find it is dead. The abdomen has already been chewed open and the meat recently dug into, but plenty remains for the taking. It tastes a little salty—like cured ham—but it's nutritious and filling. They tear off hunks, eating some but storing most of it in their social crops to bring back to the queen and the other workers. Regurgitating, they feed the queen first. For a day it's touch and go. If the queen dies, it's all over. Also, things will be difficult if all their aphid cattle die. Five are left—all in need of fresh vegetation. The strongest workers, following the evolutionary program in their genes, decide they must move their nest, including the cattle, to a more favorable locale, nearer to fresh leaves and stems. They carry the half-dead cattle eighteen feet up the beach to the long grass. Then they transport a few injured workers. Finally they carry the queen. She tucks her legs under and rolls into a limp ball—the "carrying position"—and lets herself be moved up the beach. Their new home—the dry end of a rotting log—will become the founding nest and, if they survive, a sort of command center for a program of expansion.

Over the following months the little fire ants build up strength and numbers. They may yet show the extent to which the ant wars of conquest can sometimes go. It took some years for these ants to run like conquering hordes over the Galápagos. Would the Dominican Republic be a similar battleground? The little fire ants from Costa Rica could not know that their species had already become dominant in many areas around the island. Sooner or later they would meet.

Chapter Nine

Back to the Lab

When we compare the motives which bind together the societies of humans and of ants, we are forcibly struck by their similarity.

— *Caryl P. Haskins*,
OF ANTS AND MEN, *1945*

Ants are the real sociobiology.

— *Edward O. Wilson*,
FROM A LECTURE, *1985*

It would appear that socialism really works under some circumstances. Karl Marx just had the wrong species.

— *Bert Hölldobler and Edward O. Wilson*,
JOURNEY TO THE ANTS, *1994*

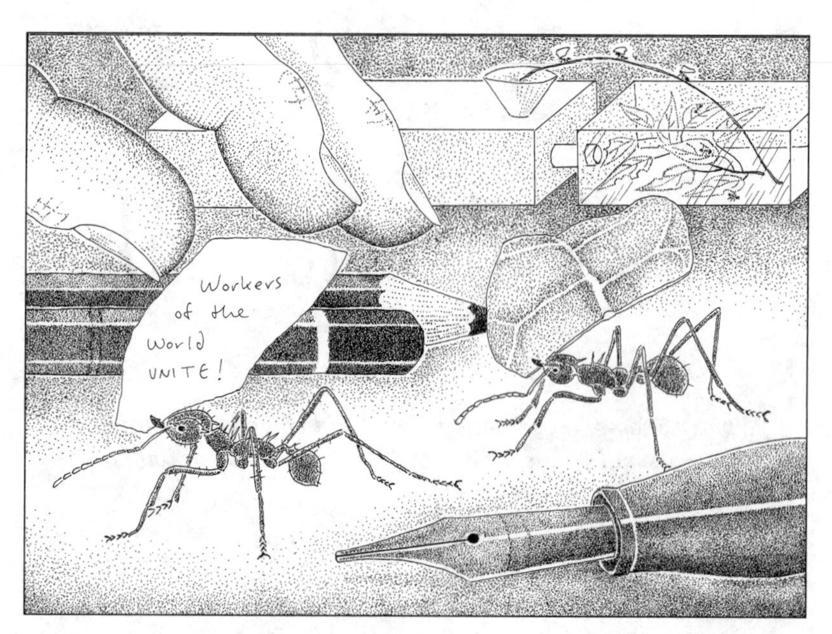

Leafcutter Ants Go to Harvard. Several colonies of leafcutter ants, Atta cephalotes, *from La Selva now live in Wilson's office at the Museum of Comparative Zoology at Harvard University. They are a source of daily wonder to Wilson and his students.*

E. O. Wilson's class on Evolutionary Biology stands as one of the most popular at Harvard. Annual "unofficial" ratings by students give high marks to the course and to Wilson for his helpful, matter-of-fact manner. Most of the class is too young to remember the sociobiology controversy, though Wilson's reputation as the father of that discipline has held sway much longer than he could have imagined. Before each class, Wilson scribbles his lecture outline on the blackboard. The points may seem strange to the students but their unfamiliarity soon yields to Wilson's lucid explanation. Dressed in tweedy brown and green with glasses, he is a distinguished member of the Harvard faculty but his down-home manner and occasional inflections make him seem sometimes a slightly awkward, small-town high school teacher. Yet his lecture turns rapidly into a one-hour roller-coaster ride of ideas, findings, and insights interspersed with anecdotes and insider asides—the best sort of teaching. Wilson is not as flamboyant or as consistently entertaining as his sometime critic, paleontologist-biologist Stephen Jay Gould, who plies his trade in the next lecture hall. Playfully esoteric by habit, master storyteller Gould earns standing ovations after some of his lectures. Wilson approaches readers and those who sit for his lectures on bended knee. His words are clear, powerful pleas for the scientific method, for rational thinking, for the conservation of the natural world, and for a mind ever open to the search for the Big Biological Picture. This last item is what has gotten Wilson into trouble in the past, and it may well do so again. Yet it has also earned him substantial praise, and a reputation that will long outlive him.

As for ants, Wilson rarely talks about them in class unless students show a strong interest. Then the stories, after class or sometimes in question period, come fast and furious, always seeming fresh, often illustrating some deeper problem in biology. Much of his work goes back to ants, but his lectures cover the wide range of his research interests present and past. His insights and his vision come from his work in diverse areas of evolutionary biology: There is the importance of modern taxonomy to try to name, classify, and study more species. There is island biogeography. There is a lecture on biodiversity, spotlighting the destruction of the forests and deserts, another on animal communication—pheromones. There is the evolution of social behavior—sociobiology—a topic he discusses in many animal groups, from coral to carpenter ants to chimpanzees. But it is his extension of sociobiology to humans—*human* sociobiology—that has landed him in controversy over the years. He's been doused with cold water, picketed by protesters, called a racist and sexist reactionary, politely ignored by his colleagues, and condemned by members of his own department. Not only his judgment was questioned, but also his ability to do good science.

He doesn't shy away from discussing it, warts and all, even with the students. In his final lecture of the year, Wilson stages an entertaining mock debate, moving from one side of the stage to the other as Wilson-Pro and Wilson-Con argue the human sociobiology question. Wilson-Pro gives many points to the "opposition" who accuse Wilson of stirring up the debate on how much genes might have influenced, if not determined, human behavior. As he attempts to see the other side, he admits, "Guilty as charged!" But his calm philosophical attitude of the 1990s stands in sharp contrast to the raw emotion that surrounded this issue two decades earlier.

As far as Wilson is concerned, the trouble started with a large book he completed in 1975 called *Sociobiology: The New Synthesis.* In it Wilson aimed to give wings to a new field of study. Part one outlines the basic concepts of social evolution, including its background rooted in population biology, group selection, and altruism. This is followed by twelve chapters constituting part two, on the various social mechanisms including communication, aggression, sex, parental care, and social symbioses. Part three, the largest section of the book, is devoted to a discussion of the "four pinnacles of social evolution": (1) colonial invertebrates—the corals, jellyfish-like siphonophores, and the bryozoans; (2) the social insects—ants, termites, and certain bees and wasps; (3) social mammals—such as wolves, lions, and apes; and (4) humans. In the book Wilson explains why he included *Homo sapiens:* "Let us now consider man in the free spirit of natural history, as though we were zoologists from another planet completing a catalogue of social species on Earth." The functional similarities between insect and vertebrate societies had long intrigued Wilson. Just as evolution by natural selection shapes the physical features of an animal, Wilson knew from his ants that evolution was shaping behavioral traits. Why not in higher animals, too? In Wilson's view, behavior that has survival value —certain types of cooperation, conflict, domination, and self-sacrifice, for example—can become encoded in the genes of any organism, including humans.

Wilson didn't invent sociobiology. The roots of sociobiology go back as far as Charles Darwin, and there have been important contributions by Konrad Lorenz and by modern evolutionary biologists in work that has focused on various organisms. But Wilson pulled the field together and systematically extended the evolutionary interpretation to the social behavior of hundreds more species. He made what seemed to him the logical decision to extend the discussion to the human species. He looked not so much at the differences between the societies of humans and animals as on the similarities.

The first notices were wonderful. Several reviewers called his book "brilliant." *The New York Times,* in a front-page account by Boyce Rensberger, heralded a new scientific discipline that "carries with it the revolutionary implication that [certain] behavior . . . may be as much a product of evolution as is the structure of the hand or the size of the brain." Overnight, Wilson's name became synonymous with sociobiology. He became its chief theorist, his book its bible. But the honeymoon was short. After scholars in the humanities questioned the implications of Wilson's grand synthesis, brutal criticism came from twenty-one Boston-area academics writing in *The New York Review of Books.* They charged that Wilson's work was "sleight of hand," his "breakthrough" the sort of biological determinism that had led to social darwinism and the racial science of the Nazi regime. The implications of Wilson's work: If human behavior is "in the genes," what happens to the notion of free will? If cultural differences between people are called genetic, does that justify the oppression of the less intelligent or less fortunate? How can sociobiology account for human cultural history—the art, music, and inventions of the creative human mind? They said it was dangerous even to suggest such theories. These critics weren't interested in debate—this was not even a subject fit for science, they declared.

Wilson was taken aback. His first reaction was to question himself: "Had I taken a fatal intellectual misstep by crossing the line into human behavior? I felt as if I had entered a nightmare world of ideology, very different from conventional biological science." Licking his wounds, Wilson started preparing a response to the so-called Sociobiology Study Group, expressing pained surprise at both the antagonistic, personal nature of the attacks and their source. Two signers had been close Harvard colleagues. Richard C. Lewontin was chairman of Wilson's own department and a professed Marxist. Wilson had helped bring him to Harvard, maintaining that Marxism was irrelevant to his scientific work. The other was Stephen Jay Gould. Three other members of the Harvard faculty were in other disciplines. Wilson faced the risk of becoming "a pariah," he reflected, "viewed as a poor scientist and a social blunderer to boot."

Few colleagues were offering even casual encouragement. Only Wilson's ant buddies—Bert Hölldobler, Bill Brown, and a few others—plus his family, supported him through the bitter parts of the controversy.

As Wilson admits in his autobiography *Naturalist*, "In 1975 I was a political naïf: I knew almost nothing about Marxism as either a political belief or a mode of analysis. I had paid little attention to the dynamism of the activist left. . . . I was not even an intellectual in the European or New York–Cambridge sense." Wilson was, if anything, a "Roosevelt liberal turned pragmatic centrist," but he was being labeled a right-wing reactionary and worse.

"The issue at hand is vigilantism," responded Wilson, in *The New York Review of Books*, "the judgment of a work of science according to whether it conforms to the political convictions of [self-appointed] judges."

Yet he knew he was out of his depth. To meet the challenge, he approached Eugene Genovese, a Marxist philosopher, to learn about Marxism, and Harvard sociologist Daniel Bell gave him a route map list of books in the fields of social sciences and the humanities. Neither liked sociobiology, yet both sympathized with Wilson because of the vicious attacks and his naïveté. How could he think that the psychologists, anthropologists, and other social scientists would drop their long-established approach to human behavior—a field they considered their own—and join an entomologist to fit their work and their discipline into some Big Biological Picture he had devised?

Wilson became a student of these and other fields new to him. He subscribed to dozens of strange magazines and journals found in few other biologists' mail boxes. But as Wilson became more astute, his critics, particularly Richard Lewontin and Stephen Jay Gould, sharpened their intellectual criticism. New radical groups, enemies of Wilson and sociobiology, emerged. He knew it would get more heated, more intense, but he never would have guessed for how long and to what extent. On bad days and good days, he longed to go back to studying ants, but for now he had something to prove. There were a number of things driving him. How did this shy, awkward myrmecologist from Alabama find himself locked in one of the most heated scientific-philosophical debates of the late twentieth century?

FOR WILSON, one strand of the thread leads back to a cross-country trip in the summer of 1952 undertaken with fellow entomologist Thomas Eisner, now a professor at Cornell University. At the time, Wilson and Eisner were first-year graduate students at Harvard eager to get on with the sort of fieldwork that real biologists do. They traced much of the perimeter of the United States, moving from biome to biome, collecting ants and other insects, living out of "Charrúa II," Eisner's 1942 Chevrolet, named after one of the native Indian warrior tribes from Uruguay. Eisner was from a German-Jewish family that had fled Hitler, moving from Spain to France to Uruguay, where Eisner had spent most of his youth, and finally to New York. Feeling a little lost and displaced from all the moving, Eisner had turned to nature, much as Wilson had done. As they collected insects together, spun tales of insect lives, and slept under the stars, they became strong friends. Somewhere driving across a dull stretch of the Great Plains, Eisner recalls, Wilson turned to him and remarked that someday, someone would have to produce a synthesis of everything that had been done on the social relations of organisms. Apparently, it was not a casual remark. The idea grew slowly.

Through the early 1950s, as Wilson worked on his Ph.D. on the common *Lasius* ants and traveled on a Harvard fellowship to the South Pacific, he started to see the diversity of ways that ants had adapted their anatomy as well as their social behavior to take the best advantage of a wide variety of ecosystems. Yet the first big step toward sociobiology came in 1959 when Wilson and his mentor Bill Brown produced their landmark paper on the "socioecology," as they dubbed it then, of the dacetine ants. It was a landmark as far as Wilson, Brown, Eisner, and a few others were concerned. Instead of the usual analysis based on physical characteristics alone, Brown and Wilson considered the social behavior, eating habits, and ecological adaptations of each species. It was one of the first studies of any social animal, large or small, to look at social behavior from an evolutionary perspective.

At the same time, the big news in biology, overshadowing every other biologist's work, was the discovery by James Watson and Francis Crick of the structure of DNA. Announced in 1953, the DNA finding came, according to Wilson, in a "lightning flash, like knowledge from the gods." It was destined to have a dramatic impact on all of biology as well as the fields of agriculture, criminal law, and business, among others. To traditional biologists who could see into the future, molecular biology held the promise of a superb tool for achieving a deeper understanding of life; instead it manifested itself as a blunt instrument that threatened most other biologists, pushing them into tiny offices or completely out of the university to make way for the new order of things. Watson, only a year older than Wilson, was then at Harvard and led the siege. Things looked bleak for people who studied large fauna—even ants were considered large fauna—or, God forbid, ecology. Non-molecular biologists were made out to be stamp collectors. "Traditional" evolutionary biologists such as Ernst Mayr were still working at Harvard, but the main impact of his excellent work was a decade or two old. The pressure was on Wilson and other young non-molecular biologists to make some serious scientific waves or be left out—not just of all the honors but of academia altogether. In fact, the solid work of Wilson and some of his contemporaries, quite apart from sociobiology, has kept "evolutionary biology" not only alive but flourishing at Harvard from the 1960s through to the 1990s.

Wilson found Watson the most unpleasant, scornful scientist he had ever met. Watson's discovery was so earth-shattering that he became a "Caligula" who could do or say no wrong. In spite of Watson's disdain, Wilson deeply admired the man's accomplishment and even his sheer audacity. He credits Watson as his "brilliant enemy" or "adverse hero." The zeal of Wilson's own future work was partly a response to the challenge of molecular biology.

Around the time Wilson and Watson were made assistant professors at Harvard in 1956, Wilson turned to communication in ants, borrowing the idea from ethology that "pieces of behavior" could be dissected and analyzed. Wilson made some key breakthroughs in pheromone research, then became interested in theoretical ecology and helped work out the first theory of biogeography. The questions were not directly pertinent to sociobiology, but the method of looking at various groups of species spurred Wilson to think on a bigger scale. By the late 1960s he began to entertain the idea of a synthesis of all the work on social insects, including wasps, ants, bees, and termites. It required massive amounts of reading outside myrmecology. To Wilson, the fresh, exciting aspect was being able to interpret all these studies from the point of view of modern population biology, Wilson's strong suit. With this overview, he brought diverse and sometimes out-of-date papers together into a coherent narrative. After eighteen months of intense work, Wilson packed *The Insect Societies* off to his publisher. Released in 1971, the book received rave reviews. It is still widely used and in print today. Even Wilson's human sociobiology critics, such as Gould, have praised the book. The book only stoked the fires of Wilson's ambition. Witness the final chapter, "The Prospects for a Unified Sociobiology," with its telling prediction, written in a bold, self-contained style:

"As my own studies have advanced, I have been increasingly impressed with the functional similarities between insect and vertebrate societies and less so with the structural differences that seem, at first glance, to constitute such an immense gulf between them. . . . The formulation of a theory of sociobiology offers one of the great manageable tasks of biology for the next twenty or so years."

As he broadened his reading, Wilson was surprised to discover that the research on vertebrate behavior was accessible to an invertebrate specialist. Maybe it's time to do that grand synthesis of all social behavior, he thought, and at the same time make sociobiology into its own scientific discipline. Says Bill Brown, "It was a big challenge. It appealed to Wilson's need to be a scout, a visionary—a role he likes to play whenever he gets a whiff of a new field opening up." It also meant that Wilson could search out the general theory he had proposed as a major task of biology. The task was somewhat easier because, for the social insects, Wilson could build on what he had already written for *The Insect Societies.* Instead of taking twenty years or so, a couple of years of eighty-hour weeks did the trick.

The reaction to *Sociobiology: The New Synthesis* taught Wilson he would have to take his case to a much wider audience than his scientific peers. He would have to sharpen his theoretical base, but at the same time he knew he had to learn how to write popular science. Several months after *Sociobiology*'s publication, he wrote "Human Decency Is Animal," an effort to explain human sociobiology in *The New York Times Magazine.* Encouraged by the result, he soon began work on his first book for a general audience, *On Human Nature,* a book that would explore human sociobiology and its implications. Meanwhile, popular articles by other writers played up the genetic determinism angle, using titles and cover headlines such as "Why You Do What You Do" *(Time),* and "It's All in the Genes" *(Discover).* A year after *Sociobiology* was published, more serious criticism began coming in from books such as *The Use and Abuse of Biology: An Anthropological Critique of Sociobiology* by University of Chicago cultural anthropologist Marshall Sahlins, who stated flatly that human behavior ought to be excluded from sociobiology. But *The Selfish Gene* by the British evolutionary biologist Richard Dawkins showed that other biologists were thinking along the same lines as Wilson. There were some honors, too, notably in 1977 when Wilson received the President's Medal for Science from Jimmy Carter. Fueled by this recognition and with *On Human Nature* about to be published, Wilson decided to undertake his most public appearance to date. The annual meeting of the "AAAS," the American Association for the Advancement of Science, one of the largest scientific gatherings of the year, attracts journalists, teachers, curious citizens, and most of all scientists from every conceivable discipline. This is the time when anthropologists, zoologists, geologists, sociologists, psychologists, geneticists, evolutionary biologists, and assorted others can creatively bump into one another, catch up on what is happening in other areas, and even at times share the platform to debate and discuss subjects of general concern.

This diverse group packed a large meeting room at the Sheraton-Park Hotel in Washington, D.C., in mid-February 1978. They were to consider a new area of interest and speculation, the emerging field of sociobiology, and they would hear from its leading proponent, Harvard professor Edward O. Wilson. It was a special two-day symposium, the kind of event that in the 150-plus-year history of the AAAS is held only when there has been either a controversy or a scientific development of some magnitude. Sociobiology qualified on both counts. For more than two years Wilson had been kept busy defending sociobiology. He was ready, even eager, for the big debate. As twenty scientists took center stage one by one to present their widely divergent views, the tension and excitement mounted. The audience was eager to hear from Wilson, who usually contends with his innate shyness with a conscious attempt to be pleasantly gregarious. But for this event he was handicapped by a broken ankle, the result of a jogging accident. He sat off to the side, nodding and smiling to associates, concentrating as he listened to each presentation while awaiting his turn. He was to be the final speaker.

Through the afternoon of his presentation, the normally sedate AAAS conference had been interrupted by demonstrators from a group of mostly young men and women calling themselves the International Committee Against Racism. Wilson had seen it all before, though on a smaller scale, and he was getting used to it. As the moderator introduced Wilson, fifteen ICAR protesters rushed onto the stage shouting, "Racist! Sexist! Fascist!" A placard saying "Racist and Fascist Scientist of the Year" was flung at Wilson. From the audience the shouts began to mount—some pro, some con, and some "let's get on with it!" Sitting at the center table on the dais, his lower leg stretched out and clearly in a cast, Wilson nervously fingered his notes. As he began to speak, one ICAR protester rushed to the table and dumped a jug of water over Wilson's head.

The audience shouted the protesters off the stage, but it took the moderator several minutes to bring the meeting back to order. He apologized to Wilson, and the audience gave the dripping scientist a standing ovation. Ever the professional, Wilson began speaking and for the next twenty minutes methodically delivered his paper. The protesters were allowed to rejoin the audience and speak their mind. No police or security guards were called. In the debate that followed, Stephen Jay Gould called the ICAR actions infantile and inappropriate. He and others distanced themselves from the demonstrators yet voiced strong comments about Wilson's "new synthesis." They respected Wilson's contributions to evolutionary biology and his work with ants, yet the idea of human sociobiology following naturally from an examination of insect and animal sociobiology was too much. And how could Wilson have the audacity to use the subtitle "the new synthesis," a clear reference to the Darwinian revolution of the 1930s and 1940s, referred to as "the modern synthesis"? To paraphrase Gould, Wilson wanted to be the greatest thing since Darwin and Freud in terms of reaching a new understanding of human nature.

In his soft-spoken voice, Wilson patiently set forth his and sociobiology's intentions. Contrary to critics' views, he said, sociobiology is *not* a "theory that human behavior has a genetic basis." Sociobiological theory allows for both possibilities: that the human brain has been "freed" from genes and that all behavior is determined by culture, or that genes influence at least some forms of social behavior. Thus, anywhere from zero to 100 percent of human social behavior could have a genetic basis. (Wilson's private guess, as of 1995, is that about 10 percent is genetic.) Says Wilson, "Genetic determinism is a perilous concept in the realm of the social sciences. To those who so earnestly desire to reject sociobiology out of hand, [genetic determinism] means that development is insectlike, proceeding down one biochemical pathway from a given set of genes to the corresponding predestined pattern of behavior. But the development of behavior can be under substantial genetic influence and still be circuitous and difficult to predict." For Wilson, science is a search for answers; political and philosophical implications are important, but one must not be hobbled by them. To make any breakthrough, he says, it is necessary to set forth theories that can be refined, tested through experiments and observation, then corrected or, if necessary, discarded.

Wilson's *On Human Nature* went on to win the Pulitzer Prize for general nonfiction. Well written, the book introduced his ideas to many more people. As the title suggests, he focused on human sociobiology, arguing eloquently that all forms of mental activity and social behavior, such as aggression, ethics, sex roles, and even religion, can be meaningfully understood against a biological background. Sociobiology, he suggested, might teach us a great deal about human nature. It was a thought-provoking book—not meant to be definitive proof. As much admiration as it earned, it also attracted more criticism.

Battle weary, Wilson was ready to go back to the ants, but a young physicist–theoretical biologist from the University of Toronto, Charles Lumsden, proposed a collaboration to try to quantify the links between genetic and cultural evolution. In 1981, Wilson and Lumsden published *Genes, Mind, and Culture,* followed, a few years later by a popular version entitled *Promethean Fire.* Says Wilson, "We produced the first theory. We took what is known about cognitive psychology, development, and so on, and tied this together in a series of models and conceptual schemes. We proposed that the genes prescribe a set of biological processes, which we call epigenetic rules, that direct the assembly of the mind. Genes are indeed linked to culture, but in a deep and subtle manner. Our model is an attempt to account for the ways in which genes and the cultural attributes of a society have evolved in parallel."

Some critics, ready and waiting, objected to having to plow through the computations when they were sure Wilson had again missed the mark. The exercise convinced none of the skeptics. Among other things, Wilson was being called a reductionist. Wilson did believe that human behavior—in the same way as animal behavior—could be broken down and studied bit by bit to try to illuminate the whole. But more than anything he was just saying it was worth a try. And with *Genes, Mind, and Culture,* he was saying, "Here's a start; let's work on it." In fact, Wilson is both reductionist and wholist, and he believes both approaches are needed to make advances in science. Gould, Lewontin, and others are too philosophically opposed to consider Wilson's approach, but even some who are sympathetic to Wilson's hypothesis have been critical. Attempting to answer his critics in a 1983 BBC roundtable discussion, Wilson was persuasive if not absolutely convincing. The moderator, John Maddox, editor of *Nature,* concluded that Wilson and Lumsden "had taken hold of an important and interesting problem, but . . . a great deal remains to be done before [their coevolution theory of genes and culture] will be comprehensively established."

Soon afterward, Wilson returned to his ants with renewed passion. He had promised his wife not to fly until their daughter was grown and had restricted his field trips mainly to long train journeys to Florida. He could work on trains and could find some interesting ants in the United States, but he needed the holy grail of the tropics with its diversity of unstudied, undiscovered species. The Brazilian Amazon, for all its magnificence, was difficult to get in and out of. In 1984, visiting La Selva, Costa Rica, with Bert Hölldobler, he knew he had found the ideal spot where he could cheerfully end his days as a myrmecologist.

Wilson began going to La Selva once every year or two, partly as a refuge from the storm. Certainly, he had made a name for himself with sociobiology, but his solid contributions in myrmecology and ecology were virtually unknown to the public. If he was known for ants, it was "the ant man who had turned to that funny idea of sociobiology." By 1991 everything had changed. Wilson had returned to the ants and had worked with passion, lightning speed, and remarkable success on a book on biological diversity, *The Diversity of Life,* as well as on many studies culminating in his opus with Bert Hölldobler on a single group of animals, *The Ants.* Their collaboration brought together almost two centuries of ant research published in a wide variety of journals, reports, and books in many languages. By the early 1990s, Wilson had effectively become once again the ant man, with the noteworthy underlying addition of being considered a passionate rain forest crusader as well. But even as Wilson wandered the paths of La Selva in 1991, he was beginning to wonder what to do next. The fierce fires of ambition were still burning.

Where did such an ambitious, apparently tireless scientific agenda come from? What are the roots and why is there so much passion? To what degree is it a personal agenda? Indeed, how much of Wilson's drive and ambition is environmental, how much genetic? Wilson explored both explanations in his 1994 autobiography, but, in his own case, he leans decidedly toward the environmental. As the first member of his family to attend university and the first and only scientist, Wilson would seem largely a self-made man. He *is,* but not to diminish his accomplishment, there are tantalizing clues in his childhood—some of it purely accidental; some of it his southern upbringing—things that happened to him that led him to the ants and, beyond the ants, to humans and to the heart of human nature.

TO ED WILSON the boy, as to E. O. Wilson the scientist, the woods will forever be a refuge from the battles of the world, a welcome place to explore and discover the infinitely fascinating, ever nurturing natural world. Early on, he says, this made him seek a career as a naturalist. But he could easily have studied snakes or perhaps even larger fauna. The same summer Wilson's parents split up, when he was seven and away enjoying camp, however, he had the career-forging injury to his eye that led him to study insects.

At age sixteen, in 1945, a year before entering the University of Alabama, Wilson decided it was time to get serious about specializing in a group of insects. He first chose flies. They were everywhere. The common housefly had given the group a bad name, but there were many others in this acrobatic group that could make for an interesting career. Attempting to order the equipment for fly collecting, he learned that the special long black insect pins, then made in Czechoslovakia, were unavailable due to World War II. Wilson turned to his second choice, the ants. They could be collected in rubbing alcohol and stored in medicine bottles. With money saved from his paper route, he bought a copy of William M. Wheeler's 1910 classic *Ants: Their Structure, Development and Behavior* and followed the author's suggestions to make glass observation cases. He has never regretted his decision. Neither flies nor snakes would have led to sociobiology. In fact, few species but ants could have stimulated visions of grand evolutionary syntheses and unified biological theories. These are the provinces of ants. Sociobiology and much of Wilson's later work depend utterly on his decision to study the social ants. Even other social species—wolves, whales, lions, and so forth—might well have led to little more outside these species. It takes much more time to study and achieve publishable results about the social behavior of large animals than with ants.

Accidents led Wilson to the ants, but that cannot account for his search for larger questions and his particular zeal in pushing human sociobiology. Studying ants alone is insufficient to get a myrmecologist to dream big. The roots of Wilson's particular mission and clues to the personal passion behind it can be found in his southern boyhood. Wilson points to his "formation" as a boy scout and the influence of military school and the southern code of honor. Certainly these have influenced his style, his method, and his manner, perhaps even formed them, but the source of his mission and his message go back to his growing up as a fundamentalist Southern Baptist. Wilson spends an entire chapter in his autobiography on his religious initiation. At age fifteen, at a revival meeting in Pensacola, Florida, Wilson followed family tradition, decided to take the step forward, to become "born again," and later experienced baptism by full immersion. It was a sincere decision, guided a little, perhaps, by the continuing contact with his real mother, who was religious, and partly in reaction to his father, who had problems with alcohol. He says he was never swayed by the rhetoric but was impressed with the drama of it all. He did feel a sense of religious awe, something he experienced many years later while hearing the all-black Baptist choir singing at Harvard in honor of Martin Luther King. His conversion was more than just "blind emotional acceptance"; there was a measure of true earnestness, as there is about everything else he has done. The next year, however, as a University of Alabama freshman, Wilson discovered Darwin and "the most beautiful theory in the world": the theory of natural selection. Here was something that could account for all the diversity of plants and animals. To Wilson, it beat religion for its sheer epic power and, more important, its ability to explain all that Wilson cared about: life on Earth. Wilson turned rebellious and began reading essays by Philip Wylie that soon stripped bare his Protestant fundamentalism. He began to feel duped by religion and decided that he had been swayed by all the theatrical staging. He toyed with Lysenkoism, enjoying the way *Heredity and Its Variability* challenged conventional science and showed that anyone could proceed to new realms of discovery, perhaps even an ant fanatic from Alabama. Having cultivated a taste for scientific rebellion, he never looked back. Wilson embraced the gospel according to Charles Darwin. Wilson's creed became evolutionary biology. He still refers to the tropical rain forest as a cathedral, the place where the biologist makes pilgrimages, goes to worship and gape in wonder at the full flowering of evolution, the place where life is more diverse than anywhere else on Earth. For him it remains a rich, apt metaphor. Today, Wilson considers himself a secular humanist, but he is really a passionate agnostic determined to search for the highest biological truth. He still has a religious-style calling, but his vocation is anti-conventional religion. Knowing the power of religion, he invokes biblical rhythms in his prose—but his words denounce the biblical explanation of the world.

Wilson may never be "through" with religion. There is an undercurrent to his work and writing that started with the publication of *Sociobiology* in 1975. Not content with having found a raison d'être and a career studying ants, Wilson seems to want to put Christian dogma into perspective, to prove it is wrong. It is a "scientific-cum-moral agenda," according to Ullica Segerstråle, a sociologist and historian of science who wrote her doctoral thesis at Harvard on the sociobiology controversy. Wilson's ultimate mission has become to bring the Darwinian gospel to the world and to make it encompass all of life—even human life. With such scientific knowledge, according to Segerstråle, Wilson believes that human beings might one day escape the influence of theologians, and develop increased control over their own lives. Only the unified theory can begin to eradicate the realm of meaning and ethics reserved for preachers. In the last chapter of *Sociobiology,* witness Wilson's eagerness to state the moral implications of his scientific work when he concludes, "A genetically accurate and hence completely fair code of ethics" must await further contributions *not* from theology and *not* from philosophy. The code, Wilson believes, will come from sociobiology.

Considering the work of scientists, we tend to overlook their private agendas. The motivations that drive scientists may have little to do with the "pure search" for objective truth. Sometimes these private agendas have interfered with good science, but often they have inspired and even directed and driven forward the work of great scientists. Albert Einstein's spiritual belief in the order of the universe guided his search for a special theory of relativity and, later, a general theory. Wilson is in the tradition of a long line of biologists and biological humanists—many of them Nobel Prize winners, including Konrad Lorenz, Francis Crick, and Jacques Monod—to suggest that biology or evolutionary biology might yield essential truths about human nature and the origins and purpose of life.

• • •

THE TEMPERATURE IN Wilson's office-lab–ant farm at Harvard's Museum of Comparative Zoology is decidedly warm—75 to 80 degrees Fahrenheit. Wilson and his wife, Renee, also keep the thermostat turned up in their suburban Lexington, Massachusetts, home. Ed, Renee, and the tropical ants that live in their home and in his office relish the heat. He has kept many species of ants over the years—from his boyhood acquaintances, the imported fire ants, to various leafcutter, *Pheidole,* and weaver ants, to a colony of harvester ants from Florida that flourished for seventeen years.

These days, in the consolidation period of his career, Wilson keeps only a few ants. Most prominent are the three leafcutter colonies collected at La Selva in 1991. He keeps them mainly to show students—and for inspiration. With his longtime assistant Kathleen Horton he feeds them a mix of northern leaves, including sugar maples, which they seem to like. Each colony—collected as a queen and one or two hundred workers, is kept trimmed to about sixty thousand members. In the wild they would already be reaching maturity after three or four years with some 3 million workers or more. But if allowed to grow to that size in Wilson's office, they would begin to take over the building. So Wilson and Horton keep each colony trimmed like a bonsai. The superorganism is kept down to a reasonable size, what might be thought of as a good fighting weight. Still, there are more than a quarter million ants here.

As the ants stay busy, working to their genetically imposed schedules and tasks, Wilson keeps to his in another part of the room. Before, during, and after the 1994 publication of his autobiography, *Naturalist,* and his memoir of myrmecology research with Hölldobler, *Journey to the Ants,* Wilson has been under pressure—hounded by his old friend, mentor, and colleague William L. Brown to finish their monograph of *Pheidole.* And so Wilson is poring over drawings and descriptions of *Pheidole* ants and scribbling comments in the margins. This long-in-process work tackles the world's largest genus of ants with six hundred species in the New World alone, an astonishing three hundred and fifty new to science. Wilson and Brown have a publisher for the seven-hundred-page monograph and a target completion date near the end of the century, which will be at least ten years after they began. A big part of the reason for the delay has been Wilson's workload. And this taxonomic monograph has been a project that threatened to shut down not one but two careers. If so, *Pheidole* will forever be known to myrmecologists as the last resting place of two of the greats. But Wilson is unwilling to consolidate everything, to call it a day just yet. In fact, he's working hard so he can get on to his next project, his next *big* project.

The phone rings. Wilson keeps his head down, doesn't even think of moving. The answering machine picks it up. It's not Brown. Someone wants some quotes for another important article on the rain forest and the loss of species diversity. For the two-thousandth time the *Pheidole* ants will have to wait. He returns the call and gives the journalist, on deadline, a couple of quotes.

Wilson's office-lab–ant farm has also doubled as "rain forest central," Harvard branch office, for much of the previous decade, since Wilson announced he was going public to join the fight to save the tropical forests. Rain forest central—it's hot enough to be! Wilson was a relative latecomer to what he refers to as the rain forest "mafia," the group of high-profile scientists—Peter Raven, Daniel Janzen, Thomas Eisner, Norman Myers, Jared Diamond, Paul Ehrlich, Thomas Lovejoy—who have all gone public to do what they can to promote the study and conservation of biological diversity. Together and separately, they have mobilized many of their fellow biologists and the public through books, articles, films, lectures, and conferences. Their work has been responsible for greater public and congressional attention to biodiversity issues—culminating in the acceptance of the biodiversity accord in 1994. The level of biodiversity funding has increased over the past decade for everything from tropical biology and systematics to plant examination and characterization studies that evaluate plants for medicines. For his part, Wilson serves on the board of the World Wildlife Fund as a science advisor, and his 1992 book *The Diversity of Life* effectively launched biodiversity studies throughout the United States and in other countries. And so Wilson has a ready line for Brown when he finally calls later on in the day to remind Wilson about "checking over those illustrations of *Pheidole.*" Wilson, half in jest, replies, "Bill, I'm sorry, I've been busy saving the world." Brown laughs and sympathizes even though he desperately wants Wilson to buckle down on *Pheidole* so they can "get the son of a bitch sorted out." But Wilson's thoughts often wander, and not just down rain forest paths. Recently, Wilson confessed to Brown, Hölldobler, and a few others his strong desire to try to synthesize all he has learned about ants and biological diversity and sociobiology, both animal and human, plus ethics. He wants to go down that road one more time, to search for Bigger Biological Pictures.

Returning to his ant-sized notations on the monograph, Wilson hesitates for a minute to gaze in thought at the biggest ant in the room—a sculpture of an ant holding a sign in an Iwo Jima–style pose that says, "Onward Sociobiology!" He had gone through a great deal with sociobiology, and yet he felt there was a lot more now to be said and done. John Updike, in his favorable review of Wilson's autobiography in *The New Yorker* in 1994, wrote that Wilson's " 'new synthesis' of sociobiology . . . fizzled." Updike added that it was "not only politically unfashionable but scientifically premature." Certainly in terms of the "instant breakthrough" that would redefine many branches of science from the point of view of evolutionary biology, it has not succeeded. But the process of science is rarely so instant, sweeping, neat, and cohesive. Updike failed to recognize that Wilson's attempt to fashion this unified theory has already proved extremely fruitful for scientific investigation. It may not have been a seminal work, to paraphrase the anthropologist Melvin Konner, who is sympathetic to Wilson's work, but it remains important. In his 1985 book *Vaulting Ambition,* Philip Kitcher criticized human sociobiology and its claims in righteous tones, but in the process, even though he failed to mention some of sociobiology's best evidence, he did document the new discipline's standing as an established academic field. As evidence of sociobiology's success, the American entomologist and scholar Howard Ensign Evans points to more than two hundred books and five scientific journals, four started since the late 1970s, devoted largely to the subject. Regarding animal sociobiology, writes Evans, "there is no question but that as a result of Wilson's work, research in this field has flourished. . . . As for human behavior, many social scientists and students of the humanities are now discussing sociobiology openly and without invoking the ire of their colleagues. Indeed, new lines of communication have opened between disciplines that in the past had remained within their own tight little circles."

Sociobiology—even though many scientists are reluctant to use the term—has insinuated itself into science, if not as a cohesive new synthesis, then certainly as a large and respectable area of inquiry. In 1988, Wilson taught the first class in human sociobiology at Harvard University—a landmark in itself. Still, the ultimate verdict on human sociobiology will not be in for decades. As Wilson wrote in *On Human Nature,* sociobiology "might at best explain a tiny fraction of human social behavior in a novel manner. Its full applicability will be settled only by a great deal more imaginative research by both evolutionary biologists and social scientists." By the 1990s, that process was well under way.

In 1995, Wilson was again turning from ants to a diet of reading that included the latest on human sociobiology, plus the wide net of everything from anthropology to neurobiology. He is planning to research and write a new book that will bring together, once again, biology, the social sciences, environment, and ethics.

In his autobiography Wilson identifies three truths that have guided all his work: "First, humanity is ultimately the product of biological evolution; second, the diversity of life is the cradle and greatest natural heritage of the human species; and third, philosophy and religion make little sense without taking into account these first two conceptions."

With any luck, Wilson's next synthesis will come out just as the century turns. Fittingly, it will be twenty-five years after the publication of *Sociobiology: The New Synthesis.* He would love nothing more than to fire up the next generation of researchers and to lay the groundwork for biological research well into the next century. He and Hölldobler have done that for ants. And Wilson has accomplished it on some scale with biodiversity studies. But can he do it with some further new synthesis down the once dangerous route of sociobiology? That will depend on how skilled and how persuasive he is. It may be Wilson's final grand attempt at the Big Biological Picture.

Chapter Ten

Revisited

Here while I lie beneath this walnut bough,
What care I for the Greeks or for Troy Town,
If juster battles are enacted now
Beneath the ants upon this hummock's crown?

—*Henry David Thoreau*,
A WEEK ON THE CONCORD AND MERRIMACK RIVERS, *1849*

[There is a] profound inquietude inspired by these creatures so incomparably better armed, better equipped than ourselves, these compressions of energy and activity which are our most mysterious enemies, our rivals in these latter hours, and perhaps our successors.

—*Maurice Maeterlinck*,
THE LIFE OF THE BEE, *1901*

Two-legged creatures we are supposed to love as well as we love ourselves. The four-legged, also, can come to seem pretty important. But six legs are too many from the human standpoint.

—*Joseph Wood Krutch*,
THE TWELVE SEASONS, *1949*

Shadow Boxing. Late in the day, two leafcutter scouts from different colonies approach each other warily from either end of a log.

THE shadows of late afternoon, when you can see them in the forest, turn tiny creatures into larger-than-life shadow puppets. Slanting through the trees at the edge of a clearing near the river, the sun makes an ant grow half a foot or more in stature. Big, lumbering rhinoceros beetles seem to turn into real rhinos. Anteaters become long-snouted furry monsters. And humans—biologists from the research station collecting and observing until late in the day—cast their long shadows over all.

On this evening two leafcutter scouts are working their way toward each other from opposite ends of a fallen log. They are from different colonies. The bold shadows of the two figures, which document their imminent meeting, are thrown up against the stiltlike buttressed roots of a massive kapok tree lying on its side. The two ants are as yet unaware of each other. They are both searching for suitable leaves, both working until sunset at a time when almost all members of their respective colonies have already returned to the nest for the night. Both keep their heads down, smelling, licking, tasting one leaf after another from the many plants growing on the decaying log. They are headed for confrontation.

Meeting in the middle, they both stop, then simultaneously step to one side, the same side. It is like a mirror image. They are the same size, the same caste, on the same mission. They begin waving their antennae, around and around like whips, to taste the air, to determine whether friend or foe. They are from two different colonies, so although they look the same, they smell different. Yet these two scouts are related. One scout comes from the colony that lived for many years at the edge of the forest near the overgrown plantation along the Río Puerto Viejo—the colony that had to move to higher ground during the flood some seven years earlier. The other scout is from the daughter colony, the leafcutters that settled under the cecropia tree near the new lab. Therefore, one scout's mother is the other scout's grandmother. This means they are niece and aunt.

But these relations have no meaning for the two scouts on the log. Their different smells—the only clues they have—say nothing of their near relation. Normally, these colonies are located far enough apart that they rarely meet each other. If possible, they would simply avoid each other. But the two colonies have both grown large, each with more than 5 million workers, and both are placing substantial demands on the vegetation within a large radius of the two colonies.

Related or not, the two ants are ready to fight. Is the danger large enough to recruit the soldiers and lay a massive trail? Or should the two of them fight it out alone? Darkness is coming; they need to get back to the nest. The faint lemony odor of citronellol leaks into the air—the defense compound of the leafcutters. They square off, open their jaws, and threaten each other, neither ant yet touching the other. Then the scout from the daughter colony slips, falls off the log—ironically, to safety. It would have been the end for one, maybe both of them, had they started to fight.

Skirmishes like these—waged along foraging trails and on the margins of colonies—are common in the ant world. Most of these are the ant wars familiar to every amateur or professional ant watcher. But even the apparently harmonious, fungivorous leafcutter superorganisms sometimes fight. When they begin to clash over food resources or a shortage of space, the conflicts can become severe.

The vision of the socialist paradise, the exquisite order and harmony of leafcutter and many other ants that Wheeler and other early ant researchers promoted, has undergone some revision in recent years. As more detailed research has been conducted, the inner workings of ant colonies have increasingly been laid bare. Leafcutter colonies can still be considered good examples of functioning socialism even if they don't extend their vision beyond the chemical boundaries of their own colony. But ant life cannot be considered utopian socialism. In army ants, for example, when the colonies reproduce by dividing in two, each half of the colony becomes increasingly polarized. In the process of establishing which virgin queens will get new colonies, daughters sometimes turn against mothers, and always there are a number of virgin queens, "ugly sisters," that are ostracized. Still, there are no fierce battles between the queens. Instead, each queen's robustness and ability to attract workers determine her fate. The losers are not killed; they are simply abandoned. In the all-or-nothing social world of the ant, however, this is a fate equal to death.

The desert strategies of the honeypot ants with their tournaments would seem—on the surface—more civilized and sensible than the conflicts of other ant species. In fact, they fight fiercely as young colonies both against other colonies of the same species and against each other—within their own colony. The conflicts center on who will be queen. After Hölldobler's work on the tournaments of the honeypot ants, he collaborated with graduate student Stephen Bartz to look at the newly founded honeypot nests, which are easily excavated with a small shovel. Digging up a few young colonies, he found not one queen but sometimes two, three, four, up to nine queens in each. Honeypot ants, both queens and males, engage in typical ant nuptial flights, mating, and then landing and shedding their wings, with the queens excavating the new colonies. But whereas most queens carry on by themselves, struggling alone to raise a first brood, the honeypot queens join together. During colony founding, pheromones make the queens intensely attractive to each other. Hölldobler knew that mature colonies have only one queen; how, then, do they get rid of the others?

A few months after the multiqueen colony is founded, each colony starts raiding other colonies nesting close to them, stealing their brood and honeypot workers, then slaughtering the queen or queens and defending workers. Some workers, attracted by the smell of the invaders, quickly change sides instead of fighting. The captured brood is then raised, and all the acquired workers help build the raiding- and workforce, at times fighting their own sisters to the death. The nests with the most queens tend to grow the fastest and are the most likely to win battles with other colonies. As the colony matures and grows larger, however, the queens inside each colony begin to face off: The dominant queen steps on her rival and points downward with her head, according to Hölldobler and Wilson in *Journey to the Ants.* "The subordinate crouches low and holds still. Any queen that consistently yields to others is eventually driven from the nest by the workers." When one queen begins to look like a winner, workers even abandon their own mothers just to be on the winning team.

By the time the honeypot colony turns three years old, only one queen is left. Colonies that start out with multiple queens produce more workers faster and therefore survive more often than those with only one. This, of course, poses a dilemma for every queen after her nuptial flight. According to natural selection, is it better to join a small colony with only two or three queens in which their individual role and chance of being the survivor are greater? Or is it better to join a large colony that is more likely to be successful even though each queen will have a correspondingly slimmer chance of becoming the one that eventually survives? The optimal number of initial queens, as predicted by Hölldobler and Bartz, is between two and three, which is about the same as the most common sizes of founding colonies in nature.

In these and other ant species that have multiple queens, there are often struggles within the colony. Multiple queens getting together to found a colony, called *pleometrosis,* has recently been studied in various species of ants. In the imported fire ants of the U.S. South, the workers kill the extra queens, while in other species the queens themselves, sometimes aided by workers, fight to the death. In any case, the cooperative arrangements among the queens are short-lived. In only a few species do one or more queens stay together for life. One of these is the little fire ant. The little fire ants have adopted a strategy to incorporate new queens into the colony. The result is a colony that expands rapidly, fighting off all other species of ants that stand in their way. Yet they refrain from fighting among the many queens and diffuse workers of their own colony. Thus, one of the most ferocious and expansionist of all ants accomplishes its agenda through working together in a massive extended colony with numerous queens that get along with one another.

The future of ant research is partly here in the inner workings of ant colonies. Only the bare outlines of the little fire ant's behavior are known. The work of Wilson on the castes of leafcutters—in which he took apart the pieces of the colony like a Swiss watch to see how it worked before putting it back together—and the research of Hölldobler and Wilson on queen competition within colonies are just the beginning. And there is still more to be studied in the universe of the ant: the developmental biology of the colony, the delineation and meaning of its communication signals, the orchestration of all the events in the life of a colony, and much more. The variation in behavior from one ant species to another is considerable—as different as among primates. Even by the most conservative estimates, almost 40 percent of the fifteen thousand presumed ant species are unknown to science, even by name. And many of the named have no more than thumbnail biographies. The quest for the evolutionary histories, or phylogenies, of the ants remains wide open. Tantalizing yet inconclusive analyses at the subfamily level have begun to question whether ponerines and some of the other ants are as primitive as once thought, while the subfamily of Neotropical army ants is now looking more derived. Whole worlds of wonder await those who would root among the societies of ants. And there will be many more lessons, revelations, and, no doubt, extrapolations to the study of evolution, ecology, and human behavior. Wilson's sociobiology will not be the last attempt to fashion a grand biological theory on the backs of social insects.

ON THE SOUTH coast of the Dominican Republic, several years after landing, the little fire ants have begun to spread down along the rows of coconut trees that divide the beach from the highway, killing colonies of various ants and extending their own colony. To anyone willing to brave the fierce stings, they would find what seems to be a series of colonies, each with its queen. But in fact the "colonies" are connected. There is only one supercolony here, and it is large and spreading. The next step in its expansion would be for one founding queen to jump the coast highway and confront the vastly different ecological conditions away from the sandy shore, where the palm trees give way to a second growth tropical scrub forest. Then they would meet many more species of ants. Victory would by no means be guaranteed; it would depend on what sort of ants live on the other side of the road, how many there are, and, crucially, how full and well defended the niches are.

On this tumultuous island where human revolutions and military takeovers have been frequent, ants, too, have had a colorful history. Wilson and Brown have both spent considerable time studying the ant fauna of the Dominican Republic, present and past, although neither has studied the little fire ant. No one knows when or where the first little fire ants arrived, but they are not present in the amber. Examining the Dominican amber, Wilson found thirty-eight genera of ants that lived in the Dominican Republic during the late Tertiary period, about twenty million years ago. At that time there were big bullet ants, miniature army ants, and aztec ants. The aztec ants are the ants most commonly found preserved in Dominican amber. Yet today all three of these ant genera are gone from the West Indies. In all, thirty-four of the thirty-eight genera survive somewhere in the Neotropics, one has left the Western Hemisphere altogether, and three are extinct. Some twenty-two remain on Hispaniola. Fifteen other genera have arrived on Hispaniola since amber times to colonize the island, pushing the modern ant fauna back up to thirty-seven genera. Thus, in the late Tertiary period, the Dominican Republic had a fauna more characteristic of the continental Americas, which suggests that back then the Greater Antilles, including the island of Hispaniola, may have been joined together and located much closer to the continental mainland. Still, as a large island—the second largest in the Antilles after Cuba—Hispaniola has kept a certain portion of its native fauna. This has protected it to some extent from being overrun by an opportunistic tramp species like the little fire ant, but it has been invaded by a number of other ants.

The story of what was in the amber and what's left today only hints at the history of ants and humans on this island and in the greater Caribbean since Columbus first arrived in 1492. Soon after Columbus arrived on Hispaniola, he helped establish Santo Domingo, the oldest European-founded city in the Americas and long the capital of the Dominican Republic. By the early 1500s an ant that stung like fire began attacking the settlers. This was not, however, the little fire ant. It was originally called *Formica omnivora,* meaning the ant that eats everything! Hölldobler and Wilson believe, however, that this was the so-called native fire ant *Solenopsis geminata.* As they report in *Journey to the Ants,* this fire ant became such a pest to the early colonists that it threatened the Spanish settlement of the New World. Colonies of the ant may have been spread from island to island as a stowaway in produce on the early Spanish ships, or it may have been one of those native species able to take advantage of the marginal, species-poor conditions opened up by human settlements. Whatever the cause, the fire ant nearly drove out the early colonists, particularly on Hispaniola. As a last-ditch effort, they called on Saint Saturnin, their patron saint, to protect them from the ant. The Catholic clergy, realizing the seriousness of the situation, led religious processions through the capital to try to exorcise the tiny yet fierce beast.

By the late eighteenth century the fire ant had reached plague proportions on several eastern Caribbean islands, including Martinique, Grenada, and Barbados. Grenada's government offered a £20,000 reward to anyone who could figure out how to exterminate this ant. It was never claimed.

To this day the same species of fire ants is found throughout most of the West Indies, but the ants have settled into a relatively peaceful existence, living side by side with other ants. Wilson isn't saying that the prayers of the early Spanish were answered; instead, these and other ants sometimes have periods of intensive conquest when new habitats open up for them, and they also have periods of comparatively peaceful living. The behavior of ants can change dramatically according to different ecological conditions. Some have a peculiar reproductive biology that under certain situations can be so favorable as to allow these ants to spread swiftly until they take over an area.

The little fire ants on the makeshift schooner took the hard route to get to the Dominican Republic. Usually, fire ants and other tramp species travel "first class," carried in food or timber in the storage compartments of ships where they are guaranteed swift and safe passage to a new port. That doesn't mean they get through customs. Fire ant colonies are among the most frequently detained insects. It is possible, of course, for tramp species to get the right papers. That's when a scientist or bona fide researcher takes out papers to import these or other insects they need for research. According to U.S. government regulations, this usually allows a colony temporary residence in a researcher's lab, but once the experiment is complete, the colony must be destroyed. Of course, some scientists, due to the open-endedness of their multiyear experiments or to preoccupation with their work, or even to a sense of affection toward their colonies and neglect of the law, do not destroy them. Some colonies have got loose in the labs, and eventually the exterminators had to be called in. One colony of Pharaoh's ants that slipped in from Brazil showed up in the early 1960s in the pipettes of a Harvard molecular biologist. This was the celebrated episode at Harvard's biology building termed the "revenge of the ants." When the ants started carrying radioactivity into their nests inside the walls, they had to be destroyed—a difficult task that occupied some of the best minds at Harvard. And fire ants, too, have gotten loose at Harvard and other public and private institutions with near disastrous results. As a boy, Wilson himself documented the arrival of the imported fire ants to Alabama, and through his life he has witnessed their unchecked spread through the American South as far west as Texas. At the 1988 Imported Fire Ant Conference, Georgia State Representative Henry Reaves expressed his philosophical approach to fire ants: "Down my way, we have more or less accepted them: work with them, kill a few, and get stung." Why are some ants such a problem? Along with Pharaoh's ants and fire ants, little fire ants, Argentine ants, several *Formica* ants, and one of the *Pheidole* ants, *Pheidole megacephala*, are among the most dominant and dangerous animals on Earth.

The reason these ants are so successful is due to a special method of colony founding or multiplication. Leafcutters and most other ants send out winged queens to found new colonies alone, but the little fire ants and others above employ a variation on the method used by army ants. Instead of dividing in two equal parts like army ants, new fire ant queens mate with their brothers and cousins in the home nest, then depart with a relatively small force of workers to start laying eggs on their own. But it's not so much colony founding as multiplication by budding. The result is a unicolonial society. When a species relies on budding to reproduce, that means it is polygynous—more than one mated queen lays eggs and lives in the same nest. When a queen splits off with a part of the colony, it does not entirely leave the colony. The superorganism does not actually divide or send out independent queens, it simply expands. The result is one enormous, extensive colony that spreads and spreads. And it keeps growing. Only when a piece of the colony with a mated queen is carried to a new island—to a place that is geographically separate—does an entirely new superorganism start up.

The little fire ants have employed these methods on numerous tropical shores. They have gotten into Florida and California, and even live in the heated greenhouses of Kew Gardens, England. But they are at their most destructive on pristine tropical islands with rare or endemic fauna, such as the Galápagos, where they can take advantage of sparsely populated species untested by competition. The imported fire ants have even taken Texas by storm. The Pharaoh's ants live in the walls of fashionable London addresses and in New York hospitals. The queens live only up to thirty-nine weeks, but they reproduce much faster than rabbits. This strategy has enabled these and a few other species to establish dominion over large areas to an extent few animals or insects ever manage. One large "supercolony" of unicolonial ants from the species *Formica yessensis,* located on the Island of Hokkaido, Japan, was estimated to contain 306 million workers and more than a million queens. The colony encompassed forty-five thousand interconnected nests covering 675 acres. This is the largest known animal society. And it is a single superorganisms.

According to Hölldobler and Wilson, "The adoption of unicolonialism, with its increased degree of inbreeding and reliance on budding as the principal means of reproduction, is an evolutionary step that parallels the adoption of apomixis [reproduction independent of fertilization] and vegetative reproduction in nonsocial organisms. Both permit the rapid growth of population in sparsely distributed habitats that are relatively free from competition. The extreme polygynous species have evolved unicolonial populations as a means for trading dispersibility for potential colony immortality." The little fire ants have traded the ability to fly high in nuptial flights, mate, and disperse away from the colony for the chance of having a superorganism become larger and larger with a network of queens and colonies fashioned into one giant ant colony—a super-superorganism whose life span, theoretically, approaches forever!

Budding again, the little fire ants on the beach east of Santo Domingo manage to cross the road at night. A single queen establishes herself a few feet beyond the road, above a gully at the forest edge. Weeks pass. A new worker brood hatches and begins to scavenge for food. They meet different species of ants which they overwhelm, kill, and eat.

And then the ants meet something unexpected—their own species. The little fire ants are not a native species, as Wilson notes, but may have been introduced by commerce—no one knows when. As worker examines worker, everything seems poised for confrontation. Because of geographic separation, it may already be in the process, the long road of developing into another species—that is, becoming reproductively separate. That will have to be left for the taxonomists to determine, but in any case, these ants have decided to fight it out. This fierce skirmish is only workers versus workers, but it spells the end of the expansion of the little fire ant colony from Costa Rica. On this, the second largest and one of the more populous Caribbean islands, the little fire ants have some stiff competition—not only from other colonies of their own or closely related species but from other colonies of ants, tramp species, that had come here earlier and won territory. The Dominican Republic was *not* the Galápagos, nor was it as small as other Caribbean islands such as Puerto Rico, which has suffered from little fire ant incursions. Had the roulette wheel of the wind and currents taken the little fire ants to St. Croix, Antigua, or perhaps Aruba, Curaçao, or Bonaire, or any number of even smaller islands or islets with few native ants, they could have had the place largely to themselves—establishing queendoms of little fire ants. But the aspiration toward world expansion is not completely dead. Not yet.

Ever since Columbus, European tourists in the Dominican Republic have come and gone, and in the early 1990s the level surpassed 2 million people a year, making it the number one travel destination in the Caribbean. It is therefore not surprising when a Spanish tourist collecting beach driftwood and shells as a souvenir of her trip puts a rather large piece into a plastic bag inside her hand luggage and departs for the airport bound for a flight to Spain.

Inside the bag, kept suitably moist with condensation and with plenty of good nesting material, are three little fire ant queens with enough of their worker colonies to ensure survival wherever they land. A myrmecologist could not have done a better collecting job, although he would have sealed the container properly!

Twelve hours later, the lady with the ants in her hand luggage disembarks from the plane. She thinks of nothing when the customs inspector asks her if she has any fruits, vegetables, or organic materials to declare. She is not searched, even though more of the ants have by now worked their way out of the plastic bag and are running around in her hand luggage looking for a way out or perhaps for a meal.

Several workers escape in the taxi—still she does not notice. Back home, on the coast just west of Valencia, she kicks off her shoes, puts her bag on the bed, and lies down beside it and falls fast asleep. Minutes later the little fire ants begin emerging from her bag. They are carrying their brood—eggs, larvae, and pupae. The three queens emerge as well. They have decided they must evacuate their nest. Forty minutes later, her bed by now a mass of ants, she dreams she is lying on the beach of the Dominican Republic. She feels an itch on her leg.

She wakes up and jumps out of bed and sees several ants on her bed. They never actually sting her. She looks at her bag and sees at once where the ants are coming from. It takes a few minutes before she realizes she must have carried them in her hand luggage! She dances out of bed, grabs the broom, and starts flailing at them. As she crushes a few of them, strong alarm pheromones are released and the ants begin to hold their mandibles open and poised. They are about to go on the warpath. Now she must watch out. She does not know she must get the queen to kill a colony—in this case three queens. A little later she retreats to the nearest telephone to call for pest control. Many of the workers get away, out the open door with the brood. They are sure to raise hell in her garden. By the time the exterminator arrives that afternoon, the ants would seem to have made their getaway. Still, he kills two queens in the general spraying through the house and around the patio. One queen has escaped. Multicolonial ants would need several colonies and successful nuptial flights to spread, but with unicolonial colonies such as this one, all it takes is one. For months afterward, little wars go on in her garden, and she has the dream about the ants in her bed. They are loose, and they are taking over the world. Each time she wakes up she is sure they are there, but no. With other battles to fight, they never return.

In the months after their "invasion" of Spain, the little fire ants make some headway, as in the Dominican Republic, but then are beaten back. The local ants and other fauna are too extensive, too well established, but if they can gain a foothold, then who knows what might happen if they or any of their progeny are carried to other Mediterranean islands? A beach in the Dominican Republic, a plane journey to Spain, a garden near Valencia may be just stepping stones in the much larger quest for world expansion.

FOR THE THIRD time in as many days, on the busy trails of La Selva, the leafcutter scout from the colony under the cecropia tree manages a narrow escape. It's mid-morning, and she is out scouting for new leaf supplies several hundred feet from the colony. All of a sudden the ground gives way, and she finds herself hanging over the edge of a pit, poised on the brink of death. With one tarsal claw grasping a piece of tree root and the other five dangling, she tries to pull herself up. At the same time she moves her antennae around to try to pick up some whiff of the danger. She smells meat coming from the pit below. It is a pitfall trap, baited by entomologists to attract carnivorous ants and other arthropods, but with equal success it can capture the unwary vegetarian or fungivore. At the bottom of the several-inch-deep pit is a smooth-sided cup—almost impossible to climb out of. Compared to similar traps employed by the ant lion larvae whose open mouths wait for insects to fall in and who also appreciate a diversity of insects, the pitfall trap is slow death. After struggling for several minutes, the scout manages to pull herself up and get back to work.

Besides encountering pitfall traps, the scout has nearly gotten swept up and bagged in a leaf litter sample that would have landed her in a Berlese funnel in the nearby lab. Berlese funnels have hot lamps that drive ants, mites and other arthropods out of dirt or leaves and down through a slippery funnel where they tumble into a bottle full of alcohol to become instantly preserved specimens. This happened to a number of her sisters. And one of her brothers, from her mother's recent royal brood, had been attracted on his nuptial flight to a white bedsheet placed behind an ultraviolet black light, set out to attract insects at night. Aztecs from a nearby tree had surrounded the mesmerized male on the screen and leisurely dismembered him, leaving before the entomologists came to examine their finds. The two scientific "hazards" the scout had not encountered were malaise traps—screen tents designed for daytime flying insects—and canopy fogging.

Canopy fogging, the newest method of insect collecting, started with U.S. National Museum researcher Terry Erwin's beetle work in Peru in 1982. Ground-based ants, such as leafcutters, are unaffected by this method, but the diverse ants of the canopy, largely unknown and little studied by entomologists, fall into an array of funnels laid out on sheets on the ground to be duly classified and counted. This method has been used a number of times at La Selva and at other prime tropical locales, and it has revealed an astonishing level of diversity. Unfortunately, the insects are killed, but the depletion is short-term and localized. In one study 20 percent of the insect species killed had returned after only ten days. Fogging has no impact on birds or mammals. The chemical used is from a natural plant product called pyrethrin, used in commercial kitchens and warehouses to control pests. Pyrethrin breaks down after a few hours in the sun.

To an ant or other insect, these various traps from the Arthropods of La Selva (ALAS) project, which is trying to collect as many kinds of insects as possible, would seem bad news. The biologists have no interest in collecting the ubiquitous leafcutters, and the same goes for aztec ants, army ants, and most of the twenty or so common ant species of La Selva. They're looking for the rarities. For the most part, too, they take only individuals, not whole colonies. In all of La Selva, their collecting efforts affect only a tiny portion of the insect fauna.

While the leafcutter scout goes back to collecting leaves, two biologists bend and stoop to pick up the cups full of insects from the various pit traps and to sweep and bag samples of leaf litter for the Berlese funnel. They also do some opportunistic collecting when they see unusual insects walking around. They remain alert for fresh tree falls—invitations to explore the canopy. If it's a cecropia tree, it may have only one ant species—the aztec ant—but certainly no more than five. But other trees have complex ecosystems in the canopy. Dozens of species of ants can be recovered from a single tree.

Since 1991 these and other researchers have been collecting samples of the insects at La Selva. It is part of a nationwide project in Costa Rica—the first of its kind in the world—to find, catalogue, evaluate, and characterize the biodiversity of an entire country.

This visionary program, started by the Instituto Nacional de Biodiversidad de Costa Rica (INBio), has its roots in the research of the Organization for Tropical Studies (OTS) in Costa Rica, such as Dan Janzen's efforts at La Selva and in Guanacaste province along the dry Pacific coast of Costa Rica, to evaluate the diversity of plants, insects, and animals. It has also been spurred on by Wilson, whose writing and campaigning efforts have given a much higher profile to biodiversity research. In late 1992, Janzen along with Rodrigo Gámez, director of INBio, and other colleagues were coming up with an estimate of five hundred thousand species for Costa Rica. This staggering figure for a country smaller than the state of West Virginia, smaller than Ireland, means that Costa Rica has from 1.7 to 17 percent of the 3 million to 30 million species thought to inhabit the Earth. Yet less than half of Costa Rica's biodiversity is actually named and known to science. Only 20 percent of the estimated three hundred thousand insect species of Costa Rica have been described. INBio knew from the start that its main work would be on insects.

In 1990, Jack Longino, Robert Colwell, and several other biologists proposed to the National Science Foundation an on-the-ground project to evaluate the diversity of insects at La Selva, to be sponsored by INBio and OTS. The three-year ALAS project was approved and in 1994 received an additional three-year extension.

And so the various species of leafcutters, army ants, aztecs, cork-headed ants, little fire ants, assorted dacetines, acrobat ants, cryptobiotic ants, wallpaper ants, the many carpenter and thief ants, the big bullet ants, and the many, many beetles, springtails, flies, butterflies, bees, wasps, and so forth, were going to be looked at, counted, and recognized as never before. By late 1994, as the results started to come in, the figures were outstanding. For ants, according to Longino, 350 species had been found. With many still to be evaluated and enumerated from the canopy, the total figure for La Selva may be five hundred species of ants. In the crown of one *Carapa guianensis* tree, Longino broke all the single tree records for ants with some sixty species collected.

The ALAS project—within the larger national INBio program—seeks to uncover the mysteries of tropical forests and to make them accessible. The stated aim is to confer value on biodiversity by making it commercially available and by teaching local people about it. Besides Longino and Colwell, the INBio program consults with international experts on each species group, but it also seeks to educate and involve local researchers and the public. Many are Costa Ricans working on the ground as part of a network of trained field collectors now being established across the country. These "parataxonomists" —local nature enthusiasts who are trained to do the first stage of collecting and preparing specimens—are crucial to the success of the program. Their efforts are already leading to increased appreciation and conservation of diversity, including insects. But their task is a large one. The idea of learning to love insects is still foreign to most people.

The project's emphasis on arthropods and other insects is mainly because in numbers of species they are the great unknown. The economic value of wild species that started with the human domestication of corn, rice, wheat, and potatoes, and the use of many plants as medicines and other industrial products, revolves around plants. But arthropods, too, have valuable uses including spider toxins used for enzyme work, luciferase genes from click beetles and fireflies for biotechnology research, and compounds from millipedes that may have promise for new drugs. Even more important are the coevolving relationships of most insects with plants—relationships that must be unraveled to permit applications in the fields of agriculture, pharmacology, and restoration ecology, among other areas. And to understand the process and history of evolution and its implications for ecology, researchers must look at all of the natural world. That's why evolutionary biologists and myrmecologists such as Wilson and Brown, botanists such as Peter Raven, ecologists such as Daniel Janzen, as well as many other biologists and ecologists today are working to conserve the tropical forests. Without substantial portions of original forest left intact, humans will be much poorer in the future—in knowledge, in helpful products, in appreciation of nature. As Wilson has said, "The destruction of biodiversity is the folly our descendants are least likely to forgive us."

THE MOTHER COLONY of leafcutters that survived the flood seven years ago and moved to higher ground still thrives. So does the daughter colony located beneath the cecropia tree ruled by the aztecs. Yet they have become increasingly competitive. The daughter leafcutter colony has also suffered friction in recent years from the massive, expanding aztec colony in the cecropia tree. The army ants, meantime, have periodically stormed through the area on their more or less annual visits. The southern swarm raider colony matured and divided the year after Wilson's visit following the Gulf War. The existing queen survived the division, as queens often do the first or second time around. And one of her daughters took half of her workforce. But the northern swarm raiders lost half their workforce after a bungled colony division. No male was able to find and enter the colony during its period of receptivity. That colony died. Yet other superorganisms have taken its place.

For many years to come at La Selva, these superorganisms will be living with one another and the researchers will be working all around them. A worker ant—single "cells" of the superorganism—may live only a few months to about a year at most, but queens of a number of species are known to live at least ten to thirty or more years. Leafcutter ant colonies have lived fourteen to fifteen years in the lab at Harvard and elsewhere, and they may live longer in the wild. Harvester ants from Idaho are known to live up to thirty years in the wild. Therefore, many ant superorganisms, replacing their interchangeable workers, can survive for decades—a life span that exceeds many mammals and is almost the equal of whales and humans. It may be possible to go back to La Selva in 2015 and see some of the same ant superorganisms that researchers found in the intensive myrmecological studies that began in the early 1980s with Wilson, Hölldobler, Brown, Longino, and others.

Of course, there are some ants that trade in their queens early to keep the superorganisms competitive, such as the army ants whose successful queens may live six years before they are usurped by their own daughters. For the army ants as well as the aztecs, the turnover is quicker. Yet in twenty-five years there might be only two to five generations of ants. And there are still other ant species—the multiqueen colonies of the little fire ants—whose superorganisms have devised a way to live theoretically forever. Trouble is, before they get there, something happens. The process often short-circuits; the plan dissolves.

Why don't all ants dispense with nuptial flights and outbreeding and take the road toward immortality? The reason is there is a price paid by these "ambitious" ants—a trade-off. Because they mate with ants that are closely related to them, they are in effect inbreeding. Inbreeding carries with it increased risk of sterility and death. The price of immortality, therefore, is the gradual, long-term weakening of a population. This may explain why fire ants or other dominant ants go through periods of swift advance followed by stagnation or decline in which advances are halted. Ants come close, very close, but in biology, the study of life, there is no ticket to immortality.

Ants have evolved not only to survive but to flourish almost anywhere on Earth. The triumph of ants is not in the minuscule brain of an individual ant, as E. O. Wilson put it, but in the success of these superbly adapted social organisms that have harnessed their energies to create thousands of clever niches on Earth. They have made the Earth their own in a way that no other species including humans has done. The ants, sharpening their survival tools with new and as yet undreamed of ways of living together, matched with their bellicose temperaments to outsiders, may well live on and on.

For several weeks, as the dry season gives way to hard, steady rain, there are brief skirmishes between foragers from the two leafcutter colonies. But these casualties are nothing compared to the looming threat from the aztecs. The leafcutter colony by the cecropia tree has suffered growing friction with the aztecs in recent years. Part of it is just the matter of the aztecs' dominating smells. The pungent odor of cyclic ketones grows stronger by the day, and any stroller outside the leafcutter colony cannot avoid encountering the smell, especially when the wind is unfavorable from the leafcutters' point of view. When leafcutter workers appear at the entrance, ready to go out and collect leaves, they are often driven back by the odor. The leafcutters' own lemony citronellol defense signals are subtler. Again and again soldiers are called out for no reason. To avoid the aztecs' smell, to avoid leading the foragers into needless conflict, the scout must make a detour out and around the cecropia and venture far away, risking death and costing more time. She continues to do this, but there is no question that the situation for the colony is strained.

But the conflict all but ends the day the giant cecropia tree crashes to the ground. Woodpeckers had been preying on the aztecs intensely in recent months, and the tree became riddled with holes, weakening it. The cecropia is already nearing the end of its life. For the aztecs who have so recently defeated the second to last aztec colony to reign supreme, it is a tough break. Now they will have to move, and they will be put at risk when they try to move their queen. Even if they can find a young, unoccupied cecropia tree to move into, they might never again be able to rule a mature tree.

The leafcutters—even those deep in the nest and the scout a few hundred feet out on the trail—feel the vibration from the fall, and for ten seconds everything stops in the leafcutter colony, even the egg production of the queen. Then, as things start moving again, a number of leafcutter workers explore outside; they notice a great light space opened in the jungle, and from holes and crevices in the fallen tree, a river of bodies pouring out like sap. The aztecs are on the run, evacuating the tree. Several aztecs are cornered by the big leafcutter soldiers, and with their fortress in ruins and their colony in confusion, the aztecs are easily killed. The others disperse into the jungle in search of another cecropia. With the aztecs' departure, the drain on the soldier resources of the leafcutters will be alleviated. But even more important, the light opening, this new gateway for the sun to enter the forest, is extremely good news for the two leafcutter colonies that had begun to compete with each other. Now, with the rainy season in full swing, there will be lush new growth in and around the edges of the clearing, and all within a short walking distance of the colony. The two leafcutter colonies will no longer have to trip over each other. For the next few years at least there will be plenty of leaves for all.

Returning home from distant foraging, the scout approaches the clearing as if it is a new world opening up. It is. For a few seconds she pauses in the full sun of the light opening, gazing up at the sky. Then, from behind her, a little leafcutter offers her a smell of something, a bit of information broadcast into the air, a few molecules of wisdom. After weeks of defensive citronellol messages, this one is faintly grassy, fruitlike, with a hint of naphtha. The little worker is recruiting the ant but not for battle. Waving her antennae, the scout follows the trail right behind the little ant to the front line at the edge of the leafcutter colony's roof, as the last of the aztecs disperse into the jungle. The old aztec colony is truly gone. Already the cyclic ketones—the pungent smells of the aztecs—are dispersing, disappearing. Soon, fresh new plants will sprout along the subtle, aromatic trails of the leafcutters. It is going to be a lush, productive summer.

Appendix I
A Chronology of Ants and Humans in Evolutionary and Recent Times

NOTE: All dates are approximate except in recent times.

4,500 million years ago (4.5 billion)	Origin of the Earth
3,500 million years ago (3.5 billion)	Oldest fossils show bacterium-like filaments and spheroids from Australia and Africa

Precambrian

700 million years ago	Oldest animal fossils

Cambrian Period (600–500 million years ago)

600 million years ago	Common ancestor of ants and humans

Ordovician Period (500–435 million years ago)

500 million years ago	Pre-insect arthropods such as trilobites and crustaceans become abundant in the seas
480–500 million years ago	Oldest plant and fungus fossils

Silurian Period (435–395 million years ago)

Devonian Period (395–345 million years ago)

390 million years ago	Early insects evolve from arthropods and begin appearing in the fossil record (Archaeognatha in Quebec shale)

Lower Carboniferous Period (345–310 million years ago)

Upper Carboniferous Period (310–280 million years ago)

Permian Period (280–230 million years ago)

240 million years ago	Most current insect orders appear in the fossil record by this time

Triassic Period (230–195 million years ago)

200 million years ago	First appearance of Hymenoptera, the insect order of bees, wasps, and ants

Jurassic Period (195–135 million years ago)

Cretaceous Period (135–65 million years ago)

130 million years ago	The evolution of the flowering plants begins. They diversify and spread around the world as the dominant form of vegetation on land. The coevolution of plants and social insects—especially bees and butterflies—begins
100+ million years ago	Early ants evolve from solitary wasps, according to earliest fossils (oldest fossil in South America) and walk and fly over the Earth along with early dinosaurs
80 million years ago	Primitive sphecomyrmine ants—as described by E. O. Wilson, F. M. Carpenter, and W. L. Brown in 1967—provide a link between the nonsocial aculeate wasps and modern ants

65 million years ago	By end of the Cretaceous period, dinosaurs are extinct and ants are on the increase

Tertiary Period (65–2 million years ago)

Paleocene Epoch (65–53 million years ago)

50–65 million years ago	Ants begin their adaptive radiation, diversifying into modern groups, increasing in number, and spreading to every corner of the world

Eocene Epoch (53–38 million years ago)

50+ million years ago	Ants assume their current position of world dominance; millions of years before humans and other mammals, ants begin living in complex social groups
50 million years ago	The beginning of the ant-fungus symbiosis in leafcutter and other gardening ants

Oligocene Epoch (38–23 million years ago)

35 million years ago	Ants appear in Baltic amber

Miocene Epoch (23–7 million years ago)

20 million years ago	Ants appear in Dominican amber, which reveals most species, or at least genera, still thriving today. Some, however, are no longer in the West Indies but are found only on mainland North and South America

Pliocene Epoch (7–2 million years ago)

3–8 million years ago	Land bridge forms linking North America and South America after tens of millions of years apart. Ants from North America invade South America and vice versa, leading to territorial expansions, extinctions, and new species

3–5 million years ago	Earliest known hominids *Australopithecus afarensis*

Quarternary Period (2 million years ago to Present)

Pleistocene Epoch—Glacial (2 million–11,000 years ago)

500,000 years ago	First *Homo sapiens*

Holocene Epoch—Recent (11,000 years ago–Present)

7,000–10,000 years ago	Development of agriculture in Near East, southern Mexico, northern Andes, and China; in the New World, leafcutter ants adapt quickly to new leaf sources
970 B.C.	King Solomon, observing harvester ants, says, "Go to the Ant, thou sluggard. Consider her ways, and be wise!"
1734	René Antoine Réaumur, father of the science of insect behavior, completes first of six volumes on insects. The seventh incomplete volume, on ants, was unearthed, edited, and translated by William M. Wheeler in 1926
1810	Pierre Huber is first to observe and write about the slavery of ants and many other aspects
1859	Charles Darwin publishes *On the Origin of Species* with his ideas about ant societies
1874	Auguste Forel publishes his classic study of the Swiss amazon ants and their slaves
	Thomas Belt makes some of the first in-depth observations of

	leafcutter and army ants, and theorizes on the coevolution of ants and acacia trees, discovering the "Beltian" bodies they feed on
1894	Father Erich Wasmann begins the modern study of ant guests as part of a career in myrmecology with many solid contributions
1899	William M. Wheeler meets the leafcutter ants of Texas and glimpses his lifework
1910	William M. Wheeler's *Ants: Their Structure, Development and Behavior* published, his exhaustive 663-page early summary of the ant world based on his ten years' work. It would be 80 years before anyone attempted a full-scale synthesis of ants again
1933	T. C. Schneirla publishes first detailed studies of army ants after work in Panama; beginning of a 40-year career unraveling the mysteries of army ants
1958	E. O. Wilson identifies chemical releasers in ants; discovers source and nature of odor trails
1959	W. L. Brown and E. O. Wilson publish landmark paper on the evolution of the dacetine ants—among the first papers to consider the socioecology of animals and one of the first

	sociobiological studies in that it emphasized the adaptation of social traits. Predates work on primates
1964	W. D. Hamilton develops concept of haplodiploidy to explain kin selection and altruism of ants and other social insects
1966	Daniel Janzen proves the coevolution of the mutualism between ants and acacia trees based on landmark field experiments to solve long-standing disagreement
1967	E. O. Wilson, F. M. Carpenter, and W. L. Brown, writing in *Science,* describe first ant remains from the Cretaceous period, Miocene age, at first thought to be 100 million years old, later revised to 80 million years old
1971	Publication of *The Insect Societies* by E. O. Wilson. Following Wheeler's *The Social Insects* (1928), this is the first effort to synthesize all literature on the social insects, using new approach of modern population biology
1975	E. O. Wilson publishes *Sociobiology: The New Synthesis* to mixed reviews and controversy that continues for several years. Follow-up book in 1978, *On Human Nature,* focusing on implications of human sociobiology, wins Pulitzer Prize

1976	Bert Hölldobler unravels the tournaments and slavery in the desert honeypot ants, reported in *Science*
1978	Robert W. Taylor rediscovers the living fossil ant *Nothomyrmecia macrops* in Australia and studies its primitive behavior
1979	World first alerted to tropical forest destruction and bleak prospects; would affect ant and other insect diversity
1990	Publication of *The Ants* by Bert Hölldobler and Edward O. Wilson; earns wide acclaim, including Pulitzer Prize the following year
1992	Publication of *Treatise on Invertebrate Paleontology,* lifework by Frank M. Carpenter of Harvard on insect fossils and evolution
1994	Barry Bolton of the Natural History Museum produces full identification key to all the ants of the world, incorporating changes of long-standing classifications at the subfamily level from 1992 paper by Cesare Baroni Urbani, Barry Bolton, and Philip S. Ward

Appendix 2
Societies for the Conservation and Study of Insects

Societies or groups specifically concerned with insects:

The Xerces Society
10 S.W. Ash Street
Portland, OR 97204
A nonprofit international conservation organization for invertebrates. Publishes an excellent quarterly membership magazine, *Wings: Essays on Invertebrate Conservation.*

Young Entomologists' Society, Inc.
1915 Peggy Place
Lansing, MI 48910–2553
A nonprofit international organization for young and amateur entomologists.

International conservation organizations that devote some time and funds to the study and conservation of insects:

Conservation International
1015 18th Street, N.W., Suite 1000
Washington, DC 20036

Defenders of Wildlife
1101 14th Street, N.W., Suite 1400
Washington, DC 20005

National Audubon Society
700 Broadway
New York, NY 10003

The Nature Conservancy
1815 North Lynn Street
Arlington, VA 22209

Rainforest Action Network
450 Sansome, Suite 700
San Francisco, CA 94111

Sierra Club
730 Polk Street
San Francisco, CA 94109

The Wilderness Society
900 17th Street, N.W.
Washington, D.C. 20006

Wildlife Conservation Society
185th Street and Southern Boulevard
Bronx, NY 10460

World Wildlife Fund-U.S.
1250 24th Street, N.W.
Washington, D.C. 20037

Worldwide Fund for Nature
WWF-International
Avenue du Mont Blanc
CH-1196 Gland
Switzerland

WWF-Canada
504–90 Eglinton Avenue East
Toronto, Ontario M4P 2Z7
Canada

WWF-Japan
Nihonseinei Akabanebashi Bldg.
3-1-14 Shiba
Minato-ku
Tokyo 105, Japan

WWF-United Kingdom
Panda House
Weyside Park
Godalming
Surrey GU7 1XR
England

Conservation organizations devoted to the study and conservation of insects and biodiversity in Costa Rica:

Organization for Tropical Studies, Inc. (OTS)
P.O. Box DM
Duke Station, NC 27706

Estación Biológica La Selva
Apartado 676
2050 San Pedro de Montes de Oca
Costa Rica

Glossary

Abdomen: the rear third of an insect's body; in ants, the section includes not only the gaster but the petiole and a portion of the thorax
Acacia Trees: trees from the genus *Acacia* found worldwide. In the dry tropics of Latin America, the leaves often carry bright orange pearl bodies, called Beltian bodies, on the leaf tips for food while the thorns offer housing for the acacia ants. Bull's-horn acacia trees have coevolved with acacia ants *(Pseudomyrmex ferrugineus)*.
Aculeate: having a sting. Most ants, as well as bees, paper wasps, and hornets, have stings.
Alarm Pheromone: a chemical substance put down by one ant for other members of the colony to create a state of alarm or alertness in the face of a common threat
Alitrunk: *see* thorax
Antbirds: any of some 28 species of Neotropical birds (family Formicariidae) that feed on the arthropods scattered by army ants
Antennae: two sensory appendages on the head that are responsible for receiving most of the information about an ant's world. Constantly in motion, the antennae have six to eleven segments covered in bristles (for touching) and cones (for smelling). See illustration on p. 23.
Appeasement Substance: a chemical secretion provided by a social parasite that reduces aggression
Arthropod: a member of the order Arthropoda, which includes all the insects, as well as spiders, crabs, lobsters, shrimp, and other crustaceans
Attinologist: a scientist who studies attine or gardening ants
Balsa Tree: a tropical rain forest tree *(Ochroma* species) having wood that is soft, is light in weight, and readily floats; often used as a substitute for cork
Berlese Funnel: a laboratory tool for processing insects in leaf litter for study. A hot lamp drives the ants, mites, or other arthropods out of dirt or leaves and down through a slippery funnel where they tumble into a bottle full of alcohol.
Biodiversity Studies: examination of different kinds of organisms, as well as studies into the use and maintenance of biological diversity

BIOMASS: the weight of all the individuals of a group of animals taken together; a measurement frequently used by ecologists
BIVOUAC: the assembled mass of army ant workers that functions as the nest and provides protection for the queen and the developing brood
BOOTY CACHE: the army ants' temporary storage areas for prey captured during their swarm and column raids
BROMELIADS: plants in the family Bromeliaceae, most of them epiphytes, typically with long, stiff leaves, showy bracts, and colorful flowers; *see also* epiphytes
CALLOW WORKER: a new, pale-colored adult ant with a soft, lightly pigmented exoskeleton
CARTON NEST: a spongy ant nest made from brown, waxlike "carton" material. Carton nests are constructed by aztec and certain other ant species whether they live inside living trees or outside in homemade nests attached to branches. The carton is made from plant fibers and secretions from the workers. The aztecs that make them outside the nest typically affix them to the trunks of large trees. They can be up to three feet across and more than seven feet long.
CECROPIA: a common Neotropical tree from the mulberry family *(Cecropia* species) that grows rapidly, readily colonizing light gaps, forest edges, and stream banks. Also called the trumpet tree, cecropias are popular with many animal species for their leaves and fingerlike catkins, which turn to fruits. However, certain cecropias enter into associations with aztec ants *(Azteca)* who move in and remodel the galleries in the hollows of the tree trunk, attacking any other insects that land on or near the tree.
CHITINOUS: having chitin, the tough protective substance that is the major component of insect and other arthropod exoskeletons
CICADA: any of a species of insect in the family Cicadidae, in the suborder Homoptera, famous for their ability to make loud, persistent sounds
CLOACA: the rear intestinal and genital tract opening in the abdomen of various insects
CLOSED CANOPY FOREST: mature forest with few or no tree falls and no light openings
COEVOLUTION: the evolution of two or more species because of mutual influence; ants, for example, have coevolved with plants and other insects in ways that make the relationship more effective and dependent
COLLEMBOLAN: *see* springtail
COMMENSALISM: symbiosis in which one member of a species benefits while the other neither benefits nor is harmed
CROP: the temporary holding pouch inside the gaster for food to be shared with the rest of the colony; see illustration on p. 78
CYCAD: a palmlike, cone-bearing, evergreen plant
CYCLIC KETONES: chemical compounds produced in the pygidial gland of aztec ants and used as an alarm pheromone
DEVELOPMENTAL BIOLOGY: the study of how organisms develop from conception to death, but focusing on the early stages
DUFOUR'S GLAND: an exocrine gland at the base of the sting, named after the French entomologist and army surgeon Léon Dufour; see illustration on p. 78
ECITOPHILE: an obligatory guest of one of the army ants, especially those of *Eciton*
ECOLOGY: the study of the interaction of organisms with their environment
ELAIOSOME: the specialized organ on certain seeds that attracts ants; the aril
ENDEMIC: a species native to a particular place and found only there
ENTOMOLOGY: the study of insects
EPIPHYTES: plants of various species that live on other plants; in the rain forest, many

plants station themselves high in the canopy to obtain more light and send out aerial roots in search of water and nutrients

ETHOLOGY: the study of animals in the context of their life cycle, ecology, and evolution. Ethology tries to interpret natural behavior such as courtship, social signals, nesting, and territorial behavior.

EVOLUTIONARY BIOLOGY: a broad category of biological studies that focus on evolution, including molecular evolution, ecology, biogeography, population biology, systematics, comparative anatomy, physiology, and ethology

EVOLUTIONARY HISTORY: *see* phylogeny

EXOSKELETON: the outside shell of the ant. Ants and other insects have skeletons on the outside of their bodies.

EXTRAFLORAL NECTARIES: the sugar glands in certain plants, usually located on the leaves or stem, that have evolved to attract ants and other insects helpful to the plant

FEMALE-CALLING ANTS: ant species that use the reproductive strategy of a virgin queen, who stands in one place and calls males to her by releasing sex pheromones

FORMICARY: an ant nest, also called a formicarium; can refer to a nest in the wild or, more commonly, in the laboratory

FUNGIVORE: an organism that subsists on fungi

FUNGUS GARDEN: the underground chambers created and maintained by leafcutters and other gardening ants where they grow their fungi

GASTER: the rear portion of an ant, behind the petiole; see illustration on p. 23

GENUS (plural, GENERA): a group of similar species with a recent common ancestor; the first part of every species' two-part name is the genus, hence the genus of *Atta cephalotes* is *Atta*

GONGYLIDIA: the minute, inflated tips of the gardening ants' fungus, which Alfred Möller called kohlrabi, because of its resemblance to heads of stalked cabbage; William Morton Wheeler renamed the kohlrabi *gongylidia*—the name commonly used today

GUEST: a social symbiont or organism that lives in symbiosis with another species, such as ants; *see* symbiosis

HOMOPTERAN: an insect from the order Homoptera, such as aphids, jumping plant lice, treehoppers, froghoppers or spittlebugs, whiteflies, coccids or scale insects, mealybugs, and related groups. Many homopteran species, which are generally soft-bodied, relatively defenseless insects, live in associations with certain ants as ant "cattle."

HONEYDEW: water and carbohydrate by-products, which are eliminated as liquid feces by homopterans and are eaten by insects such as ants. The honeydew "waste" contains key proteins, free amino acids, minerals, vitamins, and other substances.

HYMENAEA COURBARIL: *see* stinking toe tree

INFRABUCCAL POUCH: the chamber on the floor of the inside of the mouth that holds compacted waste materials and, in leafcutter virgin queens, the starter fungus for new colonies

INFRABUCCAL PELLETS: compacted waste materials held in a special pouch in the inside of the mouth

INTEGUMENT: outer coat, covering, or skin

ISLAND BIOGEOGRAPHY: the study of the geographic distribution of organisms on islands and habitat fragments

KIN SELECTION: the selection of genes that occurs when members of a species work for the survival of relatives—genes by descent from a common ancestor—rather than for one's own offspring, which is called individual selection. Kin selection is a key factor in the evolution of ants and other social insects such as bees and wasps and can be seen in other social species, including humans, though individual selection is more often the dominant factor in evolution.

LARVAE: the main brood stage in ants and other insects; the eggs hatch into larvae, wriggling whitish forms that eat solid food. As the larvae grow, they molt several times, and new skin forms. Larvae eventually turn into pupae, the final stage before adulthood.

LIANA: a climbing, woody, tropical vine

MAJOR WORKER: a member of the caste, or subcaste, of the largest worker ants, typically specialized for defense; soldier

MALE-AGGREGATION ANTS: ant species that use a mating pattern in which males from different nests assemble in a group before virgin queens join them

MANDIBLES: the pair of mouth parts or jaws of insects, including ants; see illustration on p. 23

MANDIBULAR GLAND: the exocrine gland in the head that secretes alarm pheromones in many ants; see illustration on p. 78

MEALYBUGS: certain homopteran insects that live in association with ants; *see* homopteran

MEDIA WORKER: a medium-size worker ant; a member of the medium-size worker caste or subcaste in ant colonies with three or more caste sizes

METAPLEURAL GLAND: antibiotic-secreting gland that is a diagnostic feature found in almost all modern ants; see illustration on p. 78

MIDDEN: food remains

MIGRATORY, OR NOMADIC, CYCLE: the phase of the activity cycle in an army ant colony when the colony forages actively for food and moves daily or frequently from one bivouac site to another. During this phase the bulk of the brood is in the animated larval stage and the queen does not lay eggs. *See* stationary, or statary, cycle (which is the opposite cycle).

MINIMA: minor worker ant; a member of the smallest worker caste or subcaste in a colony

MITES: various small to minute arachnid arthropods in the superorder of mites and ticks called Acari

MOLECULAR BIOLOGY: the study of the formation, structure, and activity of macromolecules, such as nucleic acids and proteins, and their role in cell replication, the transmission of genetic information, among other things; the study of biology at the molecular level

MÜLLERIAN BODIES: the food bodies produced on the bases of the leaf stalks of cecropia trees, easily detached and used by aztec ants

MYCELIUM: threadlike tufts of fungus

MYRMECOLOGY: the study of ants

MYRMECOPHILE: an organism that spends at least part of its life with ant colonies; literally, ant-lover

MYRMECOPHYTE: higher plant that lives in obligatory mutualistic relationship with ants

NATURAL SELECTION: the process by which genetic types in a population that are best adapted to their environment tend to survive and transmit their genetic characteristics in increasing numbers to succeeding generations, while those less adapted tend to be eliminated; the mechanism of evolution as proposed by Darwin

NECROPHORIC SUBSTANCES: chemicals released by an ant after death; when other ants smell these substances, they remove the corpse to the colony rubbish heap

NEUROBIOLOGY: the study of the nervous system

NUPTIAL FLIGHT: the mating flight of winged ant queens and males

OROPENDOLAS: tropical blackbirds whose whooping cries are some of those that most commonly wake the lowland Neotropical jungle at dawn

PARASITISM: *see* symbiosis

PEARL BODIES: tiny, single or multicelled, oil- and protein-rich structures, or food bodies, on the leaves or shoots of plants used to attract ants; Müllerian bodies and Beltian bodies are kinds of pearl bodies

PETIOLAR HORNS: the protruding tip of the petiole

PETIOLE (ANT): part of the narrow "waist" of an ant, situated between the thorax and gaster as part of the abdomen. Some ants have two segments called the petiole and the postpetiole. See illustration on p. 23.

PETIOLE (PLANT): leafstalk, the slender stalk attaching the leaf to the stem

PHEROMONES: the chemicals used by ants and other insects to communicate with each other. The pheromones are produced in the battery of exocrine glands and employed, as needed, to identify nestmates and enemies, to warn of danger, to call other ants to food or for help, among other uses. The meaning of a pheromone can depend on the quantity used and whether it is used alone or as a blend with other pheromones or in sequence. Sometimes pheromones are used in combination with visual or tactile signals to enhance or change the meaning.

PHORID FLIES: flies that belong to the family of small flies called Phoridae, generally smaller than common house flies. They bother many species of ants, and even the fiercest ponerines can be driven to distraction, ducking their heads and jumping around, to avoid the nuisance flies. They also live in association with army ants.

PHYLOGENY: the evolutionary history of a group of organisms, the family tree and relationships of the species in a group

PHYSOGASTRIC: the swollen condition of the abdomen of an ant such as the mated queen due to the enlargement of fat bodies, ovaries, or both

POISON GLAND: exocrine gland located on the rear underside of the abdomen or gaster. The poison gland is the source of the trail pheromone for leafcutter, Pharaoh's, and many other ants; it is the source of alarm pheromones and produces formic acid in carpenter ants and weaver ants, among others. See illustration on p. 78.

POISON GLAND VESICLES: *see* poison gland

PROTHORAX: the forward segment of the thorax in various insects

PUPAE: the preadult stage in ants, following the larval stage. Some ants spin silk cocoons to protect themselves in the pupal stage while others develop as naked pupae.

PYGIDIAL GLAND: one of the exocrine glands found in the gaster of an ant; sometimes contains trail pheromones or, less often, alarm pheromones

QUEEN-RECOGNITION PHEROMONES: the chemicals continuously released by the queen to indicate her presence in the colony and to attract the workers and control their reproductive activities; also called queen substances

RECRUITMENT TRAIL: a pheromone trail laid by scout workers to summon nestmates to a food find, a new nest site, or a site needing the workers' attention

RESTORATION ECOLOGY: the study of the structure and growth of plant and animal communities for the purpose of repairing and restoring damaged ecosystems

ROYAL BROOD: the reproductive caste of ants, comprised of virgin queens and males. *See* sterile worker caste (which is the opposite).

ROYAL CHAMBERS: the part of the nest—often the best protected and most carefully temperature-controlled—where the royal or sexual brood is raised

SEALING OFF: the blocking and isolation of unacceptable virgin queen army ants by their colony mates

SEX PHEROMONES: sex attractants; chemical substances, mainly glandular secretions, used by ants and other insects and animals to communicate their readiness to mate

SEXUAL BROOD: *see* royal brood
SLAVEMAKER: an ant that takes slaves
SLAVERY: the situation in which workers of a parasitic ant species raid the nest of another species, capture brood, and rear them as slaves
SOCIAL PARASITE: an ant or other insect that lives in the same nest as another species of ant or other insect and is parasitically dependent on it
SOCIAL STOMACH: *see* crop
SOCIOBIOLOGY: the study of the biological basis of social behavior in insects and animals, including all aspects of communication and social organization
SOLDIER: a member of the caste, or subcaste, of the largest worker ants, typically specialized for defense; major worker
SPERMATHECA: a whitish bladder below the rear of the gut in the gaster of queen ants where sperm is held as free-living cells for up to 15–20 years and used as needed to make female workers
SPIRACLES: the holes in an ant's thorax through which it breathes and exhales
SPRINGTAIL: a collembolan or member of the order Collembola, a soft-bodied, drably colored arthropod noted for a fork-shaped appendage, which is doubled up beneath the body and acts as an unpredictable spring. Most of the time these insects probe through the leaf litter, eating decaying plants and nematode worms, but if they are disturbed, they spring away.
STATIONARY, OR STATARY, CYCLE: the phase of the activity cycle in an army ant colony when the colony is subdued and relatively stationary. In this phase, the queen lays the eggs; and most of her brood is either eggs or pupae. *See* migratory, or nomadic, cycle (its opposite).
STERILE WORKER CASTE: the nonreproductive laboring caste of ants; *see* royal brood (the opposite)
STING: the sharp, piercing organ at the end of the gaster found in certain ants, as well as bees and wasps, which ejects poisonous secretions to kill, injure, or disable enemies; see illustration on p. 23
STINKING TOE TREE: a very tall, leguminous tropical tree. When the tree is mature, its leaves are avoided by herbivorous insects and mammals. Even leafcutter ants won't touch them, as the antifungal terpenoid in the leaves is poisonous to their fungus gardens. This tree produced much of the resin that fossilized the Dominican amber. It is also used by acacia ant virgin queens as a platform for calling males to mate.
STRANGLER FIG: an evergreen tree, *Ficus,* that starts as an epiphyte, gradually surrounding and strangling a host tree in an extensive network of aerial roots
STRIDULATION: the production of sound by rubbing one part of the body surface against another, which in certain ants and other insects has a communication function; leafcutter ants make this sound when they are buried underground
SUPERORGANISM: concept of ant colony as a single organism
SYMBIONT: an organism that lives in symbiosis with another species
SYMBIOSIS: the intimate, protracted, dependent relationship of members of one species with those of another, regardless of whether it benefits or harms either species. The three main forms of symbiosis are: commensalism, in which one species profits from the association without harming or benefiting the other; mutualism, in which both species benefit; and parasitism, in which one species benefits to the detriment of the other, but usually without causing its death.
SYMPHILE: a solitary insect or arthropod that is accepted to some extent by an insect colony of another species. A kind of symbiont, a symphile is typically licked, fed and/or taken to the host brood chambers, and is able to communicate with the host amicably.
SYSTEMATICS: the study of the diversity of life, including its origins and evolutionary

relationships; sometimes used interchangeably with taxonomy, but the term has taken on a wider meaning in biodiversity studies. *See* taxonomy.

TANDEM RUNNING: when one ant follows closely behind another, frequently touching the abdomen of the leader with its antennae. This occurs during exploration or recruitment in certain ant species.

TANK BROMELIADS: bromeliads with cistern basins that hold water and often become miniature insect-animal ecosystems; *see also* bromeliads

TARSAL CLAW: claw located at the tip of each foot, used for gripping

TAXONOMY: the classification of organisms, including the reconstruction of evolutionary history. *See* systematics.

TENT-MAKING BAT: a bat that roosts in banana trees, hanging upside down underneath the leaves. Where these little black bats have roosted, the midribs of the leaves are broken and the leaves are bent in half—like an A-frame tent. Green broken leaves indicate current nests; yellow leaves, recently abandoned nests; and brown leaves, old nests.

TERPENOID: one of several chemical compounds, including citronellal and citral, found in the mandibular glands and used by leafcutter and other ants as alarm pheromones

TERRITORIAL PHEROMONES: a chemical substance deposited on or around the nest that is colony- or species-specific and helps exclude other colonies; signature smell

THEORY OF ISLAND BIOGEOGRAPHY: various mathematical models and ideas that explain the numbers of species of organisms on islands and fragments of habitats

THEORY OF SPECIES EQUILIBRIUM: the theory that new species will arrive on an island at the same rate as old species residents go extinct. This is the central idea in the theory of island biogeography.

THORAX: the mid portion of an ant, between the head and petiole; see illustration on p. 23

THRIPS: very small insects from the order Thysanoptera. Most live on plants. Social behavior has recently been found in certain species of thrips.

TIBIAL SPUR: a comb attached to an ant's leg near the joint between the tibia and the tarsus, used to clean the antennae; see illustration on p. 23

TRAIL PHEROMONE: a chemical substance put down as a trail by one ant to be followed by another ant from the same colony

TRIBE: a group of similar genera with a common evolutionary origin; the taxonomic category between genus and subfamily

TROPHALLAXIS: the exchange of food between colony members, either through regurgitation from the crop or through defecation

TROPHIC EGG: a food egg; these are often undeveloped, nonreproductive eggs laid by the queen or sometimes workers, which are fed to the colony

VESTIGIAL: occurring as a rudimentary, nonfunctioning structure sometimes persisting from an earlier evolutionary form

VIRGIN QUEEN: unmated queen, a potential founder of a new colony

Guide to the Latin and Common Names of Ant Species, Genera, Tribes, and Subfamilies Used in This Book

Note: The tribes and subfamilies of all the ant species mentioned in this book are listed after their Latin scientific names, which are cross-referenced to the common names.

ACACIA ANTS: fierce ants that live in symbiosis with bull's-horn acacia trees; also known as bull's-horn acacia ants, one of the ants that keep homopteran cattle for their sugary honeydew secretions *(Pseudomyrmex ferrugineus)*
ACANTHOGNATHUS TELEDECTUS: a primitive dacetine ant with extremely long mandibles (Tribe Dacetonini, Subfamily Myrmicinae)
ACANTHOPONERA: one of the big ponerine ants from the fierce Ectatommini tribe that includes bullet ants (Tribe Ectatommini, Subfamily Ponerinae)
ACROBAT ANT: part of a large worldwide genus of ants with a somewhat heart-shaped gaster *(Crematogaster)*
ACROPYGA: a genus of formicine ants, one species of which carries its mealybug cattle in its mandibles during the nuptial flight. Once the ants are mated, this "dowry" from the mother colony ensures a ready food supply. (Tribe Plagiolepidini, Subfamily Formicinae.)
AFRICAN DRIVER ANT: the army ant of Africa, with colonies of up to 20 million and some of the largest ants in the world *(Dorylus nigricans)*
AMAZON ANTS: the genus of amazon ants, with five species of slavemakers, living variously in Europe, Russia, Japan, and North America *(Polyergus)*
AMBLYOPONE: a ponerine ant found in the Neotropics, the virgin queen of which stands on the ground and calls for males to come in to mate (Tribe Amblyoponini, Subfamily Ponerinae)
APTEROSTIGMA: small primitive fungus ants related to the big leafcutters *Atta cephalotes* (Tribe Attini, Subfamily Myrmicinae)

ARGENTINE ANT: one of the dominant ants found in many areas of the world; dubbed as one of the big, colonial fascist species bent on world conquest *(Iridomyrmex humilis)*
ARMY ANTS: ants that hunt in groups and change their nest site frequently; also called legionary ants. See also *Eciton burchelli, Eciton hamatum, Neivamyrmex.*
ATTA CEPHALOTES: a species of large leafcutter ants (Tribe Attini, Subfamily Myrmicinae)
ATTINES: gardening or fungus ants from the ant tribe Attini (Subfamily Myrmicinae). The leafcutters are the largest and most prominent attines.
ATTINI TRIBE: the attines or gardening ants (Subfamily Myrmicinae)
AZTEC ANT: see *Azteca*
AZTECA: the aztec ant, a fierce tree-living genus of ants, some of which have developed symbiotic relationships with cecropia trees, living inside them and defending them to the death; others build carton nests high in the rain forest canopy; one of the ants that keep homopteran cattle for their sugary honeydew secretions (Tribe Dolichoderini, Subfamily Dolichoderinae)
BALA: the bullet ant, or giant tropical ant, one of the largest, most fearless ants and the biggest day-to-day hazard of the Neotropical lowland rain forest *(Paraponera clavata)*
BASICEROS MANNI: the cryptobiotic ant; the slowest, dirtiest ants in the world (Tribe Basicerotini, Subfamily Myrmicinae)
BRACHYMYRMEX: one of the smallest ants in the world (Tribe Brachymyrmecini, Subfamily Formicinae)
BULLET ANT: *see* bala
CAMOUFLAGE ANTS: *see* cryptobiotic ant
CAMPONOTUS: a large worldwide genus of ants, including carpenter ants, a few weaver ants, and many others (Tribe Camponotini, Subfamily Formicinae)
CAMPONOTUS SAUNDERSI: an ant programmed to be a walking bomb. It has two oversize glands that run from head to gaster, each filled with poison. When cornered or attacked, the ant contracts its abdominal muscles violently, its body bursts open, and messy, poisonous secretions are sprayed all over, killing or disabling any opponent, as well as the ant itself. (Tribe Camponotini, Subfamily Formicinae.)
CARPENTER ANTS: part of the large worldwide genus of ants, *Camponotus*
COBRA ANT: one of the solitary ponerine hunters *(Pachycondyla villosa)*
COLOBOPSIS: cork-headed ants from Europe and North America whose soldiers stay in the nest and use their head and upper body as a sort of living door (Tribe Camponotini, Subfamily Formicinae)
COLUMN RAIDERS: army ants closely related to the swarm raiders but a little smaller and hunting in columns rather than swarms *(Eciton hamatum)*
CONOMYRMA BICOLOR: tiny desert ants that pick up pebbles in their mandibles and drop them down the entrance holes of the larger honeypot ants; rare example of tool use in insects; some reasearchers now call this genus *Dorymyrmex* (Tribe Dolichoderini, Subfamily Dolichoderinae)
CORK-HEADED ANTS: ants that use their heads to block nest entrances; see *Colobopsis* and *Zacryptocerus*
CREMATOGASTER: a large worldwide genus of ants with a somewhat heart-shaped gaster, including acrobat ants and carton ants, among others (Tribe Crematogastrini, Subfamily Myrmicinae)
CRYPTOBIOTIC ANT: a slow-moving Neotropical ant that uses camouflage as a survival strategy; called the slowest, dirtiest ants in the world *(Basiceros manni)*
DACETINE: an ant from the tribe Dacetonini (Subfamily Myrmicinae)

DACETON: a primitive dacetine ant (though some specialists have contended it's not a true dacetine) (Tribe Dacetonini, Subfamily Myrmicinae)
DOLICHODERINAE: tree-living subfamily of ants, sometimes called the odiferous ants. These ants have no sting or metapleural gland.
DORISIDRIS NITENS: a rare dacetine from Cuba (Tribe Dacetonini, Subfamily Myrmicinae)
DORYLINAE: the subfamily of Old World army ants
DORYLINES: The Old World army ants of the single genus subfamily Dorylinae
DORYLUS NIGRICANS: the African driver ant; the army ant of Africa, with colonies of up to 20 million and some of the largest ants in the world (Tribe Dorylini, Subfamily Dorylinae)
ECITON BURCHELLI: the swarm raiders, the large Neotropical army ants that hunt in huge swarms, called by some the greatest spectacle in the lowland rain forest (Tribe Ecitonini, Subfamily Ecitoninae)
ECITON HAMATUM: the column raiders, closely related to the swarm raiders but a little smaller and hunting in columns rather than swarms (Tribe Ecitonini, Subfamily Ecitoninae)
ECITONINAE: the subfamily of New World army ants, most of which live in the Neotropics; includes five genera *(Eciton, Labidus, Neivamyrmex, Nomamyrmex,* and *Cheliomyrmex)* in two tribes
ECITONINES: New World army ants of the subfamily Ecitoninae
ECTATOMMA TUBERCULATUM: the kelep ant, which is in the same tribe as the bullet ant and other large solitary hunters (Tribe Ectatommini, Subfamily Ponerinae)
ECTATOMMINI TRIBE: the ponerine tribe of large, fierce, mainly solitary hunters, including *Gnamptogenys, Acanthoponera,* the kelep, and the bullet ant (Subfamily Ponerinae).
FEVER ANTS: the subfamily of ants called Pseudomyrmecinae
FIRE ANT (IMPORTED): *see* imported fire ant
FIRE ANT (NATIVE): *see* native fire ant
FORELIUS PRUINOSUS: tiny desert ants that gather together and exude clouds of tear gas–like noxious chemicals to intimidate the larger honeypot ants (Tribe Dolichoderini, Subfamily Dolichoderinae)
FORMICA YESSENSIS: a formicine ant with the ultimate social system. One multi-queen supercolony in Hokkaido, Japan, contained 306 million workers and some 1.08 million queens in 45,000 interconnected nests stretched across a square mile. (Tribe Formicini, Subfamily Formicinae.)
FORMICIDAE: the overall family of ants, includes all 16 living subfamilies of ants and all species
FORMICINAE: the worldwide subfamily of formicine ants, which sting their enemies, injecting the caustic formic acid. Famous formicines include honeypot ants, tournament ants, weaver ants, and the numerous carpenter ants.
FORMICINES: ants from the subfamily Formicinae
FUNGUS ANTS: *see* Attini tribe
GARDENING ANTS: same as fungus ants or attines; *see* Attini tribe
GNAMPTOGENYS: a large worldwide genus of ponerine ants (Tribe Ectatommini, Subfamily Ponerinae)
HARVESTER ANTS: ants of various genera, mostly in the subfamily Myrmicinae, that regularly gather, store, and eat seeds. See *Messor* and *Pogonomyrmex.*
HONEYPOT ANTS: ants of the genus *Myrmecocystus* that have a specialized caste with enlarged crops to store quantities of the sugary liquid for the colony; often used to refer to the caste alone

HYMENOPTERA: the order of wasps, ants, and bees
HYPOCLINEA CUSPIDATUS: the nomad ant from Malaysia that keeps mealybug cattle, the only true nomad in the animal kingdom; some researchers now call this genus *Dolichoderus* (Tribe Dolichoderini, Subfamily Dolichoderinae)
HYPOPONERA: a large ponerine ant in the same tribe as the bullet ant and with a sting almost as nasty (Tribe Ectatommini, Subfamily Ponerinae)
IMPORTED FIRE ANT: the fire ant "imported" into the United States that has now taken over much of the American Southwest as far as Texas; dubbed as one of the big, colonial fascist species bent on world conquest *(Solenopsis invicta)*
IRIDOMYRMEX: the genus called meat ants
IRIDOMYRMEX HUMILIS: the Argentine ant, one of the dominant ants found in many areas of the world; dubbed as one of the big, colonial fascist species bent on world conquest (Tribe Dolichoderini, Subfamily Dolichoderinae)
KELEP ANTS: a species of ponerine ant that belongs to the same tribe as the bullet ant and other large solitary hunters *(Ectatomma tuberculatum)*
LEAFCUTTER ANTS: the higher attines or gardening tribe of ants, found only in the Neotropics, which include the *Atta* species, especially *Atta cephalotes* and *Atta sexdens*
LEGIONARY ANTS: *see* army ants
LEPTOTHORAX CURVISPINOSUS: one of the tiny myrmicine ants often enslaved by its close relative *Leptothorax duloticus* (Tribe Formicoxenini, Subfamily Myrmicinae)
LEPTOTHORAX DULOTICUS: a tiny slavemaker ant (Tribe Formicoxenini, Subfamily Myrmicinae)
LITTLE FIRE ANTS: a dominant tropical species that has spread to many islands both east and west of its origin in the Neotropics; dubbed as one of the big, colonial fascist species bent on world conquest *(Wasmannia auropunctata)*
MEAT ANTS: ants of the genus *Iridomyrmex*
MESSOR: the probable genus of harvester ants from the Bible that prompted Solomon to say, "Go to the ant, thou sluggard. Consider her ways and be wise." (Tribe Pheidolini, Subfamily Myrmicinae)
MINIATURE ARMY ANTS: the genus of mostly inconspicuous army ants that range from the Neotropics to the southern United States *(Neivamyrmex)*
MIRACLE ANT: the ant with bizarre, pitchfork-shaped mandibles *(Thaumatomyrmex)*
MONOMORIUM PHARAONIS: Pharaoh's ant; one of the world's dominant ants; dubbed as one of the big, colonial fascist species bent on world conquest (Tribe Solenopsidini, Subfamily Myrmicinae)
MYRMECOCYSTUS: the honey ants or honeypot ants, so called because of the honeypot caste with their enlarged, golden gasters that store liquid food as a strategy for living in the desert (Tribe Lasiini, Subfamily Formicinae)
MYRMECOCYSTUS MIMICUS: the tournament ant; a species of the honey or honeypot ants that has been found to engage in tournaments in the desert to size up other colonies (Tribe Lasiini, Subfamily Formicinae)
MYRMICINES: ants from the subfamily Myrmicinae
NATIVE FIRE ANT: *Solenopsis geminata*
NEIVAMYRMEX: the genus of miniature army ants that range from the Neotropics to the southern United States (Tribe Ecitonini, Subfamily Ecitoninae)
NOMAD ANT: the ant from Malaysia that keeps mealybug cattle, the only true nomad in the animal kingdom *(Hypoclinea cuspidatus)*
NOTHOMYRMECIA MACROPS: a large primitive ant—"a living fossil ant rediscovered" in southwestern Australia in 1977; a single species subfamily (Tribe Nothomyrmecini, Subfamily Nothomyrmeciinae)

ODONTOMACHUS: the trap jaw ants with the spring action jaws that can bounce invaders from the entrances to their nests (Tribe Ponerini, Subfamily Ponerinae)
OECOPHYLLA: the main genus of weaver ants, which use larval silk to make their nests; these bold formicines have perhaps the most complex pheromone communication system among the ants (Tribe Oecophyllini, Subfamily Formicinae)
ORECTOGNATHUS: a long-mandibled dacetine ant (Tribe Dacetonini, Subfamily Myrmicinae)
PACHYCONDYLA: ponerine genus; a typical, solitary hunter (Tribe Ponerini, Subfamily Ponerinae)
PACHYCONDYLA VILLOSA: the cobra ant
PARAPONERA CLAVATA: the bullet ant, or bala, one of the largest, most fearless ants and the biggest day-to-day hazard of the Neotropical lowland rain forest (Tribe Ectatommini, Subfamily Ponerinae)
PARASOL ANTS: *see* leafcutter ants
PAVEMENT ANTS: the ants that fight territorial wars on the U.S. East Coast *(Tetramorium caespitum)*
PHARAOH'S ANT: one of the world's dominant ants; dubbed as one of the big, colonial fascist species bent on world conquest *(Monomorium pharaonis)*
PHEIDOLE: the largest genus of ants, with some 1,000 to 2,000 or more species worldwide, 600 in the Americas (Tribe Pheidolini, Subfamily Myrmicinae)
PHEIDOLE BICORNIS: the ant that lives in close association with *Piper* plants in the Neotropics (Tribe Pheidolini, Subfamily Myrmicinae)
PHEIDOLE CEPHALICA: the Neotropical ants known for exhibiting the flood-evacuation response (Tribe Pheidolini, Subfamily Myrmicinae)
PHEIDOLE DENTATA: the woodland ant; an ant that uses preemptive attacks to keep at bay its much larger, dominant neighbor, the notorious imported fire ant (Tribe Pheidolini, Subfamily Myrmicinae)
PHEIDOLE MEGACEPHALA: one of the world's dominant ants; dubbed as one of the big, colonial fascist species bent on world conquest (Tribe Pheidolini, Subfamily Myrmicinae)
POGONOMYRMEX: this myrmicine genus includes many species of harvester ants that collect seeds and store them underground (Tribe Myrmicini, Subfamily Myrmicinae)
POLYERGUS: the genus of amazon ants, with five species of slavemakers living variously in Europe, Russia, Japan, and North America (Tribe Formicini, Subfamily Formicinae)
POLYERGUS RUFESCENS: the amazon ant of Switzerland, the best known and longest studied slavemaker, which takes slaves (various *Formica* species) on special "slave raids." The amazons are large, robust ants, mostly shiny red in color.
POLYRHACHIS: golden-haired ants with sharp hooklike spines (Tribe Camponotini, Subfamily Formicinae)
PONERINES: the largely solitary or giant hunting ants from the subfamily Ponerinae
PRIONOPELTA AMABILIS: the wallpaper ant; a primitive rain forest ponerine that wallpapers its nest for humidity control (Tribe Amblyoponini, Subfamily Ponerinae)
PROCERATIUM: one of the ponerine ants, one species of which uses its rear abdomen to block its nest from intruders (Tribe Ectatommini, Subfamily Ponerinae)
PSEUDOMYRMECINAE: the subfamily of ants sometimes called fever ants
PSEUDOMYRMEX: New World genus of acacia and other ants, mainly from the Neotropics (Tribe Pseudomyrmecini, Subfamily Pseudomyrmecinae)
PSEUDOMYRMEX FERRUGINEUS: the acacia ants; fierce ants that live in symbiosis with bull's-horn acacia trees; also known as bull's-horn acacia ants, one of the ants that keep homopteran cattle for their sugary honeydew secretions (Tribe Pseudomyrmecini, Subfamily Pseudomyrmecinae)

SOLENOPSIS: a large genus containing various fierce fire ants, as well as thief ants (Tribe Solenopsidini, Subfamily Myrmicinae)
SOLENOPSIS GEMINATA: the native fire ant (Tribe Solenopsidini, Subfamily Myrmicinae)
SOLENOPSIS INVICTA: the imported fire ant; the fire ant "imported" into the United States that has now taken over much of the American Southwest as far as Texas; dubbed as one of the big, colonial fascist species bent on world conquest (Tribe Solenopsidini, Subfamily Myrmicinae)
SPHECOMYRMA FREYI: the wasp ant, the first fossil ant to be found from the Cretaceous period (from the extinct subfamily Sphecomyrminae)
STRUMIGENYS: a long-mandibled dacetine ant (Tribe Dacetonini, Subfamily Myrmicinae)
STRUMIGENYS WILSONI: a dacetine ant (Tribe Dacetonini, Subfamily Myrmicinae)
SWARM RAIDERS: the large Neotropical army ants that hunt in huge swarms, called by some the greatest spectacle in the lowland rain forest *(Eciton burchelli)*
TETRAMORIUM CAESPITUM: the pavement ants that fight territorial wars on the U.S. East Coast (Tribe Tetramoriini, Subfamily Myrmicinae)
THAUMATOMYRMEX: the miracle ant with bizarre, pitchfork-shaped mandibles; belongs to a one-genus tribe (Tribe Thaumatomyrmecini, Subfamily Ponerinae)
THIEF ANTS: part of the large genus of ants called *Solenopsis*
TRAP JAW ANTS: the ponerine ants with the spring action jaws that can bounce invaders from the entrances to their nests *(Odontomachus)*
TOURNAMENT ANTS: a species of the honey or honeypot ants that has been found to engage in tournaments in the desert to size up other colonies *(Myrmecocystus mimicus)*
WALLPAPER ANT: a primitive rain forest ponerine that wallpapers its nest for humidity control *(Prionopelta amabilis)*
WASMANNIA AUROPUNCTATA: the little fire ants, a dominant tropical species that has spread to many islands both east and west of its origin in the Neotropics (Tribe Blepharidattini, Subfamily Myrmicinae)
WASP ANT: the first fossil ant to be found from the Cretaceous period, long extinct *(Sphecomyrma freyi)*
WEAVER ANTS: ants mainly of the genus *Oecophylla* (plus some *Camponotus* species) that use larval silk to make their nests; see *Oecophylla* and *Camponotus*
WOODLAND ANT: an ant that uses preemptive attacks to keep at bay its much larger, dominant neighbor, the notorious imported fire ant *(Pheidole dentata)*
ZACRYPTOCERUS: cork-headed ants; ants that use their heads to block nest entrances; one of the ants that keep homopteran cattle for their sugary honeydew secretions (Tribe Cephalotini, Subfamily Myrmicinae)

Bibliography

Much of this book is based on firsthand experience and detailed interviews with scientists in the field and the laboratory. I have also used the following as key sources throughout the book:

BARNES, ROBERT D. 1980. *Invertebrate Zoology.* Fifth edition. Orlando, Fla.: Harcourt Brace Jovanovich.

BARONI URBANI, CESARE, BARRY BOLTON, AND PHILIP S. WARD. 1992. "The Internal Phylogeny of Ants (Hymenoptera: Formicidae)." *Systematic Entomology* 17: 301–29.

BOLTON, BARRY. 1994. *Identification Guide to the Ant Genera of the World.* Cambridge, Mass.: Harvard University Press.

BRANDÃO, C. R. F., R. G. MARTINS-NETO, AND M. A. VULCANO. 1990. "The Earliest Known Fossil Ant (First Southern Hemisphere Mesozoic Record) (Hymenoptera: Formicidae: Myrmeciinae)," *Psyche* 96: 195–208.

BREED, M. D., J. H. FEWELL, A. J. MOORE, AND K. R. WILLIAMS. 1987. "Graded Recruitment in a Ponerine Ant." *Behavioral Ecology and Sociobiology* 20, no. 6: 407–11.

BROWN, WILLIAM L., JR., AND EDWARD O. WILSON. 1959. "The Evolution of the Dacetine Ants." *Quarterly Review of Biology* 34: 278–94.

CHAPELA, IGNACIO H., STEPHEN A. REHNER, TED R. SCHULTZ, AND ULRICH G. MUELLER. 1994. "Evolutionary History of the Symbiosis Between Fungus-Growing Ants and Their Fungi," *Science* 266: 1691–94.

EVANS, HOWARD E., ed. 1984. *Insect Biology.* Reading, Mass.: Addison-Wesley Publishing Company.

EVANS, MARY A. AND HOWARD E. 1970. *William Morton Wheeler, Biologist.* Cambridge, Mass.: Harvard University Press.

FORSYTH, ADRIAN, AND KENNETH MIYATA. 1981. *Tropical Nature.* New York: Charles Scribner's Sons.

GOTWALD, WILLIAM H., JR. 1982. "Army Ants." In *Social Insects,* edited by H. R. Hermann, vol. 4. New York: Academic Press, 157–254.

———. 1995. *Army Ants: The Biology of Social Predation.* Ithaca, N.Y.: Cornell University Press.

HINKEL, GREGORY, JAMES K. WETTERER, TED R. SCHULTZ, AND MITCHELL L. SOGIN. 1994. "Phylogeny of the Attine Ant Fungi Based on Analysis of Small Subunit Ribosomal RNA Gene Sequences," *Science* 266: 1695–97.

HOGUE, CHARLES L. 1993. *Latin American Insects and Entomology.* Berkeley and Los Angeles: University of California Press.

HÖLLDOBLER, BERT, AND EDWARD O. WILSON. 1990. *The Ants.* Cambridge, Mass.: Harvard University Press.

———. 1994. *Journey to the Ants: A Story of Scientific Exploration.* Cambridge, Mass.: Harvard University Press.

JANZEN, DANIEL H. 1966. "Coevolution of Mutualism Between Ants and Acacias in Central America." *Evolution* 20: 249–75.

———. 1967. "Interactions of the Bull's Horn Acacia (*Acacia cornigera* L.) with an Ant Inhabitant (*Pseudomyrmex ferruginea* F. Smith) in Eastern Mexico." *University of Kansas Scientific Bulletin* 47: 315–558.

———. 1969. "Allelopathy by Myrmecophytes: The Ant *Azteca* as an Allelopathic Agent of *Cecropia.*" *Ecology* 50: 147–53.

———, ed. 1983. *Costa Rican Natural History.* Chicago: University of Chicago Press.

KRICHER, JOHN C. 1989. *A Neotropical Companion: An Introduction to the Animals, Plants and Ecosystems of the New World Tropics.* Princeton, N.J.: Princeton University Press.

LONGINO, JOHN T. 1991. "*Azteca* Ants in *Cecropia* Trees: Taxonomy, Colony Structure, and Behaviour." In *Ant-plant Interactions,* edited by C. R. Huxley and D. F. Cutler. Oxford: Oxford University Press.

———. 1994. "How to Measure Arthropod Diversity in a Tropical Rainforest." *Biology International* 28: 3–13.

MCDADE, LUCINDA A., KAMALJIT S. BAWA, HENRY A. HESPENHEIDE, AND GARY S. HARTSHORN, eds. 1994. *La Selva. Ecology and Natural History of a Neotropical Rain Forest.* Chicago: University of Chicago Press.

MOFFETT, MARK. 1993. *The High Frontier: Exploring the Tropical Rainforest Canopy.* Cambridge, Mass.: Harvard University Press.

RETTENMEYER, CARL W. 1963. "Behavioral Studies of Army Ants." *University of Kansas Scientific Bulletin* 44: 281–465.

SCHNEIRLA, THEODORE C. 1971. *Army Ants: A Study in Social Organization*, edited by H. R. Topoff. San Francisco: Freeman & Co.

SUDD, JOHN H., AND NIGEL R. FRANKS. 1987. *The Behavioural Ecology of Ants.* Glasgow and London: Blackie, and New York: Chapman and Hall.

WEBER, NEAL A. 1972. *Gardening Ants, the Attines.* (Memoirs of the American Philosophical Society, vol. 92) Philadelphia: American Philosophical Society.

———. 1982. "Fungus Ants." In *Social Insects,* edited by H. R. Hermann, vol. 4. New York: Academic Press, 255–363.

WILSON, EDWARD O. 1971. *The Insect Societies.* Cambridge, Mass.: Belknap Press, Harvard University Press.

———. 1975. *Sociobiology: The New Synthesis.* Cambridge, Mass.: Belknap Press, Harvard University Press.

———. 1992. *The Diversity of Life.* Cambridge, Mass.: Belknap Press, Harvard University Press.

———. 1994. *Naturalist.* Washington, D.C.: Island Press, Shearwater Books.

WOOTTON, ANTHONY. 1984. *Insects of the World.* London: Blandford Press.

I have also drawn on some of the classics of entomology:

BELT, THOMAS. 1874. *The Naturalist in Nicaragua.* London: John Murray.

BORGMEIER, THOMAZ. 1955. "Die Wanderameisen der neotropischen Region." *Studia Entomologia* 3: 1–716.

FOREL, AUGUSTE. 1874. *Les Fourmis de la Suisse.* Zurich: Société Helvétique des Sciences Naturelles.

———. 1921–23. *Le Monde social des fourmis.* 6 vols. Translated in 1928 as *The Social World of the Ants, Compared with That of Man.* 2 vols. New York and London: Putnam.

HASKINS, CARYL P. 1945. *Of Ants and Men.* London: George Allen & Unwin.

HUBER, PIERRE. 1810. *Recherchés sur les moeurs des fourmis indigènes.* Paris: J. J. Paschoud.

MANN, WILLIAM M. 1948. *Ant Hill Odyssey.* Boston: Little, Brown.

MÖLLER, ALFRED. 1893. "Die Pilzgärten Einiger Südamerikanischer Ameisen." *Botanische Mittheilungen aus den Tropen* 6: 1–127.

WHEELER, WILLIAM M. 1910. *Ants: Their Structure, Development and Behavior.* New York: Columbia University Press.

———. 1928. *The Social Insects: Their Origin and Evolution.* New York: Harcourt, Brace & Co.

Acknowledgments

MY DEEPEST THANKS are due first and foremost to Ed Wilson for so many talks and interviews over the years and for that first invitation to discover ants in Costa Rica. I am equally indebted to Bill Brown. Besides the many wonderful stories he told while we prospected for ants, Bill and his wife, Doris, later entertained me for several days at their Ithaca, New York, home. It felt like a return to the tropics because of the hundreds of orchids hanging from trees all around their house. I am also grateful to many other myrmecologists for interviews and responses to detailed letters, including Barry Bolton (The Natural History Museum, London), Bert Hölldobler (Theodor Boveri Institut, University of Würzburg, Germany), Ted R. Schultz (Smithsonian Institution), James K. Wetterer (Wheaton College), John H. (Jack) Longino (Evergreen State College and La Selva Biological Station), and Neal A. Weber. Stefan Cover and Gary Alpert from The Ant Room at the Museum of Comparative Zoology (MCZ), Harvard University, were helpful with sources and told wonderful Bill Brown stories. John de Cuevas was a delightful partner on the first La Selva visit, and the Clarks, David and Deborah, then station directors, were gracious hosts, with Donald Stone, Charles Schnell, and Marybel Soto-Ramirez helping on behalf of the Organization for Tropical Studies. Gayle Davis-Merlen from the Estación Científica Charles Darwin on the Galápagos supplied valuable references and stories on the advance of the little fire ants. Idelisa Bonnelly de Calventi first invited me to the Dominican Republic so I could follow the story there. The enthusiasm and expertise of the late botanist Calvin R. Sperling helped give me a deep respect for plants and their key role in the lives of ants and other insects.

Bob Bender, my editor at Simon & Schuster, provided just the right kind of encouragement, understanding, and helpful guidance. I am grateful for this chance to work with him. Douglas Leiterman, John Oliphant, Robert Emmet Hoyt, and the editors of *The Atlantic* first expressed delight at the idea of getting down to ant level and following these amazing animals through their remarkable lives. Later, thanks to Michael Rosen and a period as writer-in-residence at The Thurber House, I had a chance to flesh out my approach and create some text to interest a publisher. Betty Hoyt assisted with research and sending out proposals; Patricia Donahey helped field the initial responses. Two editors, Heidi von Schreiner and Catherine Crawford, pushed the idea along considerably with their initial suggestions and encouragement, for which I am grateful. Ted Schultz, in between finishing at Cornell University and moving to the Smithsonian Institution, checked the final manuscript, and it benefited considerably from his sharp eye and eclectic background. To prepare the illustrations, Ruth Pollitt took the characters of the book to heart and helped make them come alive. I am indebted to Victor McElheny and the Knight Science Journalism (former Vannevar Bush) Program at MIT for expanding my understanding of how science works and plays in the minds of its practitioners. Finally, to one such practitioner, my wife, Sarah Wedden, I would like to express particular gratitude for our shared life of children, science, music, writing, and much more.

Index